fMRI

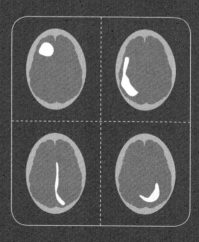

fMRI

PETER BANDETTINI

The MIT Press | Cambridge, Massachusetts | London, England

This book was set in Chaparral Pro by Toppan Best-set Premedia Limited.

Library of Congress Cataloging-in-Publication Data

Names: Bandettini, P. A. (Peter A.), author.
Title: fMRI / Peter A. Bandettini.
Description: Cambridge, MA : MIT Press, [2020] | Series: The MIT Press
   essential knowledge series | Includes bibliographical references and index.
Identifiers: LCCN 2019013953 | ISBN 9780262538039 (paperback : alk.
   paper)
Subjects: LCSH: Brain—Magnetic resonance imaging. | Magnetic resonance
   imaging.
Classification: LCC RC386.6.M34 B36 2020 | DDC 616.8/047548—dc23
LC record available at https://lccn.loc.gov/2019013953

CONTENTS

SERIES FOREWORD

The MIT Press Essential Knowledge series offers accessible, concise, beautifully produced pocket-size books on topics of current interest. Written by leading thinkers, the books in this series deliver expert overviews of subjects that range from the cultural and the historical to the scientific and the technical.

In today's era of instant information gratification, we have ready access to opinions, rationalizations, and superficial descriptions. Much harder to come by is the foundational knowledge that informs a principled understanding of the world. Essential Knowledge books fill that need. Synthesizing specialized subject matter for nonspecialists and engaging critical topics through fundamentals, each of these compact volumes offers readers a point of access to complex ideas.

Bruce Tidor
Professor of Biological Engineering and Computer Science
Massachusetts Institute of Technology

In taking the first step and picking up this book, you may be wondering if this is just another book on fMRI (functional magnetic resonance imaging). To answer: This is not just another book on fMRI. While it contains all the basics and some of the more interesting advanced methods and concepts, it is imbued, for better or worse, with my unique perspective on the field. I was fortunate to be in the right place at the right time when fMRI first began. I was a graduate student at the Medical College of Wisconsin looking for a project. Thanks in large part to Eric Wong, my brilliant fellow graduate student who had just developed, for his own non-fMRI purposes, the hardware and pulse sequences essential to fMRI, and my co-advisors Scott Hinks and Jim Hyde who gave me quite a bit of latitude to find my own project, we were ready to perform fMRI before the first results were publicly presented by the Massachusetts General Hospital group on August 12, 1991, at the Society for Magnetic Resonance Meeting in San Francisco. After that meeting, I started doing fMRI, and in less than a month I saw my motor cortex light up when I tapped my fingers. As a graduate student, it was a mind-blowingly exciting time—to say the least. My PhD thesis was on fMRI contrast mechanisms, models, paradigms, and processing methods. I've been developing and using fMRI ever since. Since 1999, I have been at the National Institute of Mental Health, as chief of the Section on Functional Imaging Methods and director of the Functional MRI Core Facility that services over thirty principle investigators. This facility has grown to five scanners—one 7T and four 3Ts.

Thousands of researchers in the United States and elsewhere are fortunate that the National Institutes of Health (NIH) has provided generous support for fMRI development and applications continuously over the past quarter century. The technique has given us an unprecedented window into human brain activation and connectivity in healthy and clinical populations. However, fMRI still has quite a long way to go toward making impactful clinical inroads and yielding deep insights into the functional organization and computational mechanisms of the brain. It also has a long way to go from group comparisons to robust individual classifications.

The field is fortunate because in 1996, fMRI capability (high-speed gradients and time-series echo planar imaging) became available on standard clinical scanners. The thriving clinical MRI market supported and launched fMRI into its explosive adoption worldwide. Now an fMRI-capable scanner was in just about every hospital and likely had quite a bit of cheap free time for a research team to jump on late at night or on a weekend to put a subject in the scanner and have them view a flashing checkerboard or tap their fingers.

Many cognitive neuroscientists changed their career paths entirely in order to embrace this new noninvasive, relatively fast, sensitive, and whole-brain method for mapping human brain function. Clinicians took notice, as did neuroscientists working primarily with animal models using more invasive techniques. It looked like fMRI had potential. The blood oxygen level–dependent (BOLD) signal change was simply magic. It just worked—every time. That 5% signal change started revealing, at an explosive rate, what our brains were doing during an ever-growing variety and number of tasks and stimuli, and then during "rest."

Since the exciting beginnings of fMRI, the field has grown in different ways. The acquisition and processing methods have become more sophisticated, standardized, and robust. The applications have moved from group comparisons where blobs were compared—simple cartography—to the machine learning analysis of massive data sets that are able to draw out subtle individual differences in connectivity between individuals. In the end, it's still cartography because we are far from looking at neuronal activity directly, but we are getting much better at gleaning ever more subtle and useful information from the details of the spatial and temporal patterns of the signal change. While things are getting more standardized and stable on one level, elsewhere there is a growing amount of innovation and creativity, especially in the realm of post-processing. The field is just starting to tap into the fields of machine learning, network science, and big data processing.

The perspective I bring to this book is similar to that of many who have been on the front lines of fMRI methodology research—testing new processing approaches and new pulse sequences, tweaking something here or there, trying to quantify the information and minimize the noise and variability, attempting to squeeze every last bit of interesting information from the time series—and still working to get rid of those large vessel effects!

This book reflects my perspective of fMRI as a physicist and neuroscientist who is constantly thinking about how to make fMRI better—easier, more informative, and more powerful. I attempt to cover all the essential details fully but without getting bogged down in jargon and complex concepts. I talk about trade-offs—those between resolution and time and sensitivity, between field strength and image quality, between specificity and ease of use.

I also dwell a bit on the major milestones—the start of resting state fMRI, the use and development of event-related fMRI, the ability to image columns and layers, the emergence of functional connectivity imaging and machine learning approaches—as reflecting on these is informative and entertaining. As a firsthand participant and witness to the emergence of these milestones, I aim to provide a nuanced historical context to match the science.

A major part of fMRI is the challenge to activate the brain in just the right way so that functional information can be extracted by the appropriate processing approach against the backdrop of many imperfectly known sources of variability. My favorite papers are those with clever paradigm designs tailored to novel processing approaches that result in exciting findings that open up vistas of possibilities. Chapter 6 covers paradigm designs, and I keep the content at a general level: after learning the basics of scanning and acquisition, learning the art of paradigm design is a fundamental part of doing fMRI well. Chapter 7 on fMRI processing ties in with chapter 6 and again, is kept at a general level in order to provide perspective and appreciation without going into too much detail.

Chapter 8 presents an overview of the controversies and challenges that have faced the field as it has advanced. I outline twenty-six of them, but there are many more. Functional MRI has had its share of misunderstandings, nonreproducible findings, and false starts. Many are not fully resolved. As someone who has dealt with all of these situations firsthand, I believe that they mark how the field progresses—one challenge, one controversy at a time. Someone makes a claim that catalyzes subsequent research, which then either confirms, advances, or nullifies it. This is a healthy process in such a dynamic research climate, helping to focus the field.

This book took me two years longer to write than I originally anticipated. I appreciate the patience of the publisher Robert Prior of MIT Press who was always very encouraging. I also thank my lab members for their constant stimulation, productivity, and positive perspective. Lastly, I want to thank my wife and three boys for putting up with my long blocks of time ensconced in my office at home, struggling to put words on the screen.

I hope you enjoy this book. It offers a succinct overview of fMRI against the backdrop of how it began and has developed and—even more important—where it may be going.

INTRODUCTION

Every so often, methodological breakthroughs come along that open entirely new vistas of unexplored scientific questions, directions of study, and potential applications. Functional MRI (fMRI) is such a breakthrough. In 1991, fMRI was more discovered than developed, in that the basic technology and methodology were already available to a handful of centers. The realization that this technology could enable observation of subtle, localized hemodynamic changes associated with brain activation was all that was needed. This noninvasive, relatively high-speed and high-sensitivity method to map human brain activity was able to spread rapidly thanks to the abundance of fMRI-capable existing clinical MRI scanners already in existence. Functional MRI also precipitated a significant change in the landscape of neuroscience research. Within a short time, thousands of scientists around the world not only embraced fMRI as a new and powerful method to complement their ongoing studies but also went on to redirect their entire research focus on this revolutionary technique.

Before the arrival of fMRI, cognitive neuroscientists were limited to probing behavioral responses or measuring signatures of brain activity using three techniques. Electroencephalography (EEG) and magnetoencephalography (MEG) measure respectively electrical and magnetic signals on the scalp. Positron emission tomography (PET) scanners detect the spatial location of injected tracers labeled with ionizing radiation, to map flow as well as metabolism. While EEG is relatively inexpensive, MEG and PET scanners are not, so their use was limited. Scalp-based measures suffer from low spatial resolution and a certain degree of uncertainty as they rely on source localization models that must address the "inverse problem" in that different sources may provide the same measured fields at the scalp. Measures from PET have generally lower spatial resolution and temporal resolution than fMRI, and do not allow many repeated experiments due to the invasiveness of injecting radioactive substances in the body.

Many neuroscientists who previously focused on animal models gave fMRI a try. With the emergence of fMRI, a wide range of neuroscientists from many subdisciplines converged on this method as they increasingly recognized it offered either highly complementary or simply superior information for their studies. Following a period of

Within a short time, thousands of scientists around the world not only embraced fMRI as a new and powerful method to complement their ongoing studies but also went on to redirect their entire research focus on this revolutionary technique.

growth, fMRI began to fill large spatial and temporal gaps of our knowledge of the brain's functional organization—exploring the systems-level organization of human and nonhuman brains.

While fMRI is a thriving technique today—with almost five thousand papers published per year and a total of almost sixty thousand papers published since 1992, it is still growing in many ways. The number of groups using fMRI continue to increase. The level of sophistication in data acquisition and processing has advanced from simple subtraction approaches to machine- and deep-learning approaches on massive multi-subject databases, and by all measures is accelerating in its development. The depth and certainty with which we can interpret the signal continues to improve with many studies comparing fMRI to more direct neuronal measures as well as physiologic parameters. More subtle neuronal and physiologic information is still being squeezed from the signal. Finally, massive databases of fMRI data sets are being created, curated, and analyzed. With all its achievements, fMRI has only a few solid clinical applications. This is a major shortcoming and the reasons include difficulty in streamlining its implementation in the clinical setting, lack of high functional contrast, and noise and variance at the individual subject level. The differences in functional brain organization between normal and clinical populations are subtle and the variance within groups is large, thus impeding, for now, single-subject

Following a period of growth, fMRI began to fill large spatial and temporal gaps of our knowledge of the brain's functionalorganization—exploring the systems-level organization of human and nonhuman brains.

assessments from fMRI data. Clinical implementation also will likely need to leverage the ability for fMRI to provide real-time information to either the clinician to guide the scan or the subject to provide meaningful brain activity feedback for therapeutic purposes.

This book is about fMRI—with an emphasis on fMRI methods. It starts with a brief overview of brain imaging methodologies (chapter 2), then delves into an overview of fMRI signal (chapter 3), fMRI contrast (chapter 4), MRI acquisition (chapter 5), paradigm design (chapter 6), and processing methods (chapter 7). Chapter 8 covers the contentious issues that have helped advance the fMRI field.

My intention is to provide useful basic information as well as perspective on the field. Care has been taken to minimize jargon, eliminate equations, and to explain all the essential concepts as clearly as possible, with some strategically placed redundancy. This book aims to be equally helpful, informative, and entertaining to the graduate student, the established research scientist, the curious reader, and the clinician. My academic career essentially started with fMRI, and it's what I've been working to develop ever since that moment on September 14, 1991, when I first saw the MRI signal from the motor cortex—my own motor cortex—deviate upward during a finger-tapping task. It's a pleasure to share with readers some of the collective knowledge from the field.

THE OTHER HUMAN BRAIN
IMAGING METHODS

To understand the impact and significance of fMRI it's important to put the brain imaging methodology landscape into perspective, highlighting what the other techniques can and cannot do. This summary is not comprehensive— but highlights what is possible without fMRI, and suggests why fMRI, when it arrived, filled such a large niche.

Before the inception of brain imaging, the functional significance of specific brain areas could only be inferred by determining behavior deficits associated with localized brain injuries. For example, in 1861, Paul Broca identified regions in the brain associated with word production from studying his patient who had severe damage in the posterior of the left inferior frontal gyrus and could produce only the syllable "tan." In 1874, Carl Wernicke identified the area important to forming and understanding coherent sentences, based on his patient who had damage to the boundary of the temporal and parietal lobes that impaired understanding and production of speech and writing. In 1909, Tatsuji Inouye made the first rough human retinotopic maps from studying visual deficits in patients who had bullet wound-induced lesions in the visual cortex. Perhaps the most famous lesion study was that of Phineas Gage,[1] who, while working on a railroad in 1848, suffered damage to his left frontal lobe when a tamping iron was launched by an accidental explosion through the front left side of his head—specifically the left orbital frontal cortex. He survived and mostly recovered and had no immediately obvious deficits. However, his personality was altered. The implications of his injuries are still debated today. The lesion-based approach has led to these and other insights into the functional significance of brain regions in the human brain, but the ultimate impact of human lesion mapping limited how generalizable the findings are as well as the detail with which such functional localization could be determined. Lesioned areas may be found essential for multiple functions or may not be essential at all to compromised behavior observed. With structural and functional imaging, our understanding of the brain advanced beyond the reliance on damaged brains to inform how a healthy brain functioned.

With structural and functional imaging, our understanding of the brain advanced beyond the reliance on damaged brains to inform how a healthy brain functioned.

Structural Imaging

Our ability to image brain anatomy in healthy humans began in 1895 with the invention of the first X-ray-based radiograph by Wilhelm Roentgen. This method is well suited for imaging bones, which have a density sufficient to attenuate X-rays, but is less optimal for imaging and differentiating soft tissue in the brain. In about 1961, computed tomography (CT) imaging began with the work of William Oldendorf. CT was able to form volumetric images, however it still relied on X-rays. Magnetic resonance imaging (MRI) was introduced in the mid-1970s when Paul Lauterbur published the first images using magnetic resonance imaging (MRI).[2] Lauterbur and Peter Mansfield were awarded the Nobel Prize in 2003 for inventing, respectively, "slice selection"—a clever use of magnetic field gradients to localize a slice in space—and "echo planar imaging" or EPI—a high-speed MRI approach that is the foundation of most functional MRI. For imaging the brain, MRI was superior to CT as it could image and differentiate a wide range of soft tissue types. Because of this unique advantage of MRI, along with its noninvasiveness, MRI scanner use exploded worldwide, starting in the early 1980s, quickly becoming an indispensable clinical tool. This explosion in utility later helped launch fMRI as most MRI scanners could easily be converted into fMRI scanners.

MRI is a powerful and diverse technology, allowing characterization of tissue types as well as detection of the presence of lesions. In the brain, MR images highlighting gray matter, white matter, fat, cerebrospinal fluid (CSF), tumors, trauma, hemorrhage, fiber tract connections, iron concentration, blood flow, blood oxygenation, electrical conductivity, macroscopic molecule content, elasticity, and much more can be created.

Tissue-specific MRI parameters that allow differentiation include longitudinal relaxation (T1), transverse relaxation (T2 and T2*), proton density (So), density of macromolecules, flow velocity characteristics, magnetic susceptibility, and diffusion coefficient. MRI systems, like computers, are controlled by programs that determine when the various fundamental components (such as radio frequency excitation and magnetic field gradient direction and amplitude) of each scan are played out. These programs are called pulse sequences and can be adjusted to highlight any of the various parameters just mentioned, allowing for a large and still-growing number of possible contrasts. The variety of anatomic contrasts have steadily increased as MRI scanners have advanced. Improvements in MRI have included the use of exogenous contrast agents to further enhance lesion-specific contrast, the use of magnetic susceptibility contrast to highlight blood-brain barrier disruptions, and the use of diffusion-weighted sequences to highlight inflammation and edema. Lastly,

MRI resolution, speed of acquisition, and image quality have been improving since the first grainy low-resolution scans were produced in the late 1970s.

Diffusion is the random thermal motion of small particles such as water molecules in a given medium. The effects of molecular diffusion on MR signals have been studied since the 1950s. A significant improvement of diffusion measurement using MR techniques was made in the 1960s by utilizing magnetic gradient pulses that lead to a decrease in signal in proportional to the diffusion coefficient of the water molecules. Diffusion-weighted imaging (DWI) was developed in the 1980s.[3] In unrestricted free diffusion, the diffusivity is uniform along all directions, or isotropic. However, a diffusion process in biological tissue, such as brain white matter, tends to be anisotropic (having a certain directional preference) because the diffusive molecules typically experiences direction-dependent restrictions due to arrangements of tissue structures. Diffusion tensor imaging (DTI) was developed in the 1990s[4] as a tool to quantify the anisotropy of diffusion in biological tissue. Tractography, a promising technique to delineate myelinated neuronal pathways based on DTI or beyond-tensor techniques, has also been developed.[5]

DTI and tractography have been used to create dramatic images of white matter tracts in individual subjects, allowing the inference of region-to-region connectivity. In recent years, the concept of brain connectivity has been developed such that inferences about connectivity variations associated with developmental diseases, traumatic brain injury, and even normal variance among individuals—all based on DTI-derived connectivity maps—have been made. The use of DTI for connectivity mapping is relatively new and promises to be an essential part of the tool chest available to clinicians for more precise individual assessment. Toward the goal of better understanding brain organization, new insights about brain connectivity structure across different scales have been put forward using DTI.[6] An outstanding recent book highlighting all things related to diffusion MRI is titled, aptly, *Diffusion MRI*, edited by Derek Jones.[7]

Hemodynamic and Metabolic Assessment

The development of *functional* brain imaging followed a somewhat parallel time line to that of the previously described anatomic brain imaging. In the 1880s Angelo Mosso conjured an approach to hypothetically infer the change in blood flow with increased brain activity known as the "human circulation balance" method. Subjects would lay supine on a balance table, and on them being presented with a stimulus or a demanding cognitive task, the table would tilt toward the head as cerebral blood volume increased in the brain. A detailed account of Mosso and his experiments is found in the book *Angelo Mosso's Circulation of Blood in the Human Brain*, edited by Marcus Raichle and Gordon Shepherd.[8]

After a few decades of relative quiescence, other techniques began to emerge to provide the first blood-flow maps of the brain. Quantitative measurement of cerebral blood flow became possible when S. S. Kety and C. F. Schmidt introduced their "Kety-Schmidt" method in 1945 by which a chemically inert indicator gas such as nitrous oxide was equilibrated in the brain.[9] The time-dependent difference between venous and arterial concentrations of the indicator gas as it was reaching equilibrium was found to be proportional to cerebral blood flow.

The "Xenon-inhalation" method for brain imaging was developed in the 1960s and 1970s by Neils Lassen, David Ingvar, and Erik Skinhoj, and involved subjects inhaling a radioactive tracer: Xenon-133.[10] This tracer infiltrated the blood, and the spatial distribution of radioactivity and thus flow was measured using appropriately placed radioactivity detectors—known as scintillators. Later, up to 254 scintillators allowed two-dimensional images of brain activity to be produced on a color monitor, revealing the first human functional brain images. In the late 1970s and early 1980s these now iconic images showed regional changes in flow associated with speaking, reading, visual or auditory perception, and voluntary movement and were a source of inspiration to aspiring brain imagers—including myself—worldwide.

The development of radioligands jumpstarted the functional imaging revolution in the 1980s. Radioligands either remain within the blood stream or cross the blood-brain barrier to bind with receptors. Radioligands are either single-photon or positron emitters. Methods for measuring the radioactivity from these radioligands include single-photon emission computed tomography (SPECT) and positron emission tomography (PET). The first human PET scanner was developed in the 1970s. Functional imaging was further advanced by the development of oxygen-15-labeled water imaging. H_2O^{15} emits positrons and creates images where the signal strength is proportional to regional cerebral blood flow. Since brain activity leads to regional increases in blood flow and blood volume, H_2O^{15} PET, measuring flow, allowed investigators to create maps of brain activation or the increase in blood flow associated with cognitive tasks. Later, a more common functional imaging method that was based on PET, used fluorodeoxyglucose (FDG), a positron-emitting sugar derivative that is distributed in the brain according to local metabolic activity. Unlike the short half-life of O^{15} (2.25 minutes), the 110-minute half-life of FDG F^{18} allowed PET scans to be performed by devices that were physically distant from the cyclotron producing the isotope, opening up a much wider utility.

In the 1970s researchers discovered that near-infrared (NIR) light was extremely useful for imaging hemodynamic changes with brain activation. The brain is semitransparent to NIR light in the range of 700–900 nm.[11] The key to the functional imaging utility of this technique are the unique oxygen-sensitive light absorption properties of hemoglobin-enabling near-infrared spectroscopy (NIRS), a useful method for assessing noninvasively human brain activation on the cortical surface just beneath the skull. Differences in the absorption spectra of deoxy-Hb and oxy-Hb enable the measurement of relative changes in hemoglobin concentration as well as total blood volume using light attenuation at multiple wavelengths; thus any activation-induced changes in blood volume or oxygenation occurring near the surface of the brain can be mapped and quantified.

NIR spectroscopy methodology was further developed by pioneers such as Britton Chance,[12] and it is a thriving method complementary to other functional imaging techniques. Vendors have recently produced wearable NIR devices, allowing many uses—including bedside hemodynamic assessment and brain-machine interfacing. Functional NIR spectroscopy (fNIRS) relies on the now well-established understanding that localized brain activation is accompanied by an increase in blood flow, volume, and oxygenation in activated regions. Blood oxygenation increases because blood delivery exceeds the oxygen uptake by activated neurons. This overabundance of oxygen delivery to activated tissue is also the basis of the fMRI signal. Studies comparing fMRI and fNIRS show similar spatial patterns and timings of activation-induced hemodynamic responses. Functional NIRS has several advantages over fMRI in terms of cost and portability but cannot be used to measure cortical activity more than a few centimeters deep due to limitations in light-emitter power. If using fNIRS through the skull, functional resolution is also less than that of fMRI due to the distance of the probes from the brain.

Electrophysiologic Assessment and Imaging

A few decades prior to the inception of metabolic and hemodynamic methods for assessing and mapping brain activity, methods for assessment and mapping of the electrical activity of the brain were being developed. In 1924, Hans Berger recorded the first human electroencephalogram (EEG), and by 1938, electroencephalography had gained widespread acceptance.

EEG is a method that allows detection of electrical activity in the brain. It is noninvasive, with the electrodes placed on the scalp. Neuronal membranes typically hold a net negative membrane potential of −70 millivolts (mV). Depolarization occurs at the postsynaptic membrane in activated neurons and is understood to be the primary mechanism by which neurons transfer information. Repolarization occurs following depolarization and involves reestablishing the membrane potential. EEG measures transient electrical dipoles generated by the local summation of electrical current flow across cellular membranes during depolarization as well as during the reestablishment of electrical equilibrium. EEG most often is used to diagnose epilepsy, sleep disorders, coma, and brain death.

EEG has limited spatial resolution as well as limited certainty in the localization of brain activity because different electrical conductance of brain tissue, cerebrospinal fluid, and skull tends to add ambiguity when attempts are made to calculate the neuronal activity sources. In addition to the ambiguity introduced by different conductance of different tissues, the problem of localization of activity is intrinsically "ill posed" as there are many solutions (hypothetical dipole sources) that could create the measured electrical activity at the scalp. Despite these limitations, EEG continues to be a valuable tool for research and diagnosis, especially when millisecond-range temporal resolution is required. It also has the advantage of being a direct measure of neuronal activity—not relying on hemodynamic changes.

A derivative of the EEG technique is known as event-related potential (ERP) measurement. Measurement of evoked potentials involves averaging the EEG activity time-locked to the presentation of a brief stimulus or task. Event-related potentials (ERPs) refer to averaged EEG responses that are time-locked to more complex processing of stimuli; this technique is used in cognitive science research to temporally disentangle specific cognitive processes that occur on the order of milliseconds to hundreds of milliseconds.

Rather than being sensitive to *electrical* signals on the scalp as with EEG, magnetoencephalography (MEG) is sensitive to subtle magnetic fields that are constantly fluctuating on the scalp due to brain activity from hypothetically more focal sources in the brain. Basic physics dictates that any electrical current produces a magnetic field around the current carrying element. An advantage of MEG over EEG is that magnetic fields generated by the brain and their manifestation on the surface of the skull are not affected by the inhomogeneity of the electrical conductivity of the brain and skull, therefore allowing more precise source localization than EEG of neuronal current sources. Because of this lack of distortion, there is certainty in estimating the location of activity. However, the inverse problem still exists in that many potential sources of current combinations can give rise to the same magnetic field pattern on the surface of the skull.

The signal measured by both techniques is mostly from the activity of pyramidal neurons, which constitute about

70% of the cells in the cortex. These pyramidal neurons are oriented perpendicular to the cortical sheath. While EEG records electrical activity oriented perpendicular to the surface of the brain, MEG records activity oriented parallel to the surface. Depending on the relative orientation of the sheath to the surface of the skull, the different techniques have different sensitivities.

In the 1980s, MEG manufacturers began to arrange multiple sensors into arrays to cover a larger area of the head. Present-day MEG arrays are set in a helmet-shaped dewar that typically contain 300 or more sensors, covering most of the head. In this way, MEG scans can now be accumulated rapidly and efficiently. Cutting-edge MEG technology development has created sensors that do not require superconducting wires, and therefore are much more mobile, allowing subjects to wear a MEG system on their head.[13] This may open up an entirely new area of brain mapping as new paradigms involving the subject moving naturally in the world are developed.

Applications of MEG include basic research into rapidly changing perceptual and cognitive brain processes, localizing regions affected by pathology before surgical removal, determining the function of various parts of the brain, and, recently, in carrying out neurofeedback studies.

Magnetic Resonance Spectroscopy

MRS (magnetic resonance spectroscopy) is a noninvasive technique that is based on nuclear magnetic resonance (NMR). Chemists and physicists use MRS to analyze and characterize small molecules in solid, liquid, and gel-like solutions. Unlike MRI, which measures the signal from protons in abundant water molecules having a specific resonance frequency, MRS usually detects the signal from compounds, each characterized by a unique resonant frequency, in much lower concentrations. Magnetic resonance spectroscopy can be used to detect nuclei that include carbon (13C), lithium (7Li), nitrogen (15N), fluorine (19F), sodium (23Na), phosphorus (31P), and hydrogen (1H). However, only the phosphorus and hydrogen are present in sufficient abundance to be detected in humans. Hydrogen is the most commonly detected nucleus due to its high natural abundance and easily discernible spectra.

The types of biochemicals (metabolites) that can be studied include choline-containing compounds (which are used to make cell membranes), creatine (a chemical involved in energy metabolism), inositol and glucose (both sugars), N-acetyl aspartate, alanine, and lactate (which are elevated in some tumors), and neurotransmitters such as glutamate and gamma-Aminobutyric acid (GABA). Glutamate and GABA are particularly important as they are known to be neurotransmitters associated with excitatory and inhibitory brain activity, respectively. Alterations in these are associated with several psychological and neurological disorders.

The first in vivo 31P MRS experiment was conducted on a mouse head using a conventional spectrometer.[14] Spectra of brain without contamination from other tissues were first obtained using a localization technique with a surface coil.[15] In past decades, MRS has evolved into a useful tool for biological research and clinical diagnosis. The lack of sensitivity of MRS is the primary reason why spectroscopy and spectroscopic imaging have not broken into regular clinical use. In addition, MRS is extremely sensitive to magnetic field imperfections and other system instabilities that do not as adversely affect MRI in general.

At present, MRS is mainly used as a tool for medical research, but it has potential for providing useful clinical information. Better coils, novel pulse sequences, higher field strengths, and more stable scanners—enabling increased sensitivity—may someday push MRS into more common clinical practice. MRS currently is used to investigate several diseases in the human body, most notably cancer (in brain, breast, and prostate), epilepsy, Alzheimer's disease, Parkinson's disease, and Huntington's disease. It has also been used to assess tissue damage with traumatic brain injury.

fMRI

Functional magnetic resonance imaging (fMRI) has completely altered the landscape of human systems-level brain research. The technique has key advantages that have precipitated swift adoption by the brain research community. Five are listed here. First, fMRI can be carried out on clinical scanners, and hospitals worldwide have MRI scanners capable of fMRI. This first point cannot be overemphasized. If, for instance, fMRI required the mass production of an entirely new piece of technology that had very limited clinical efficacy—thus limiting access to a much larger market—fMRI's growth would have been severely stunted as scanners would be much more expensive and such investments prohibitive for most neuroscientists. The use of fMRI on clinical scanners enabled rapid growth using the already thriving clinical scanner infrastructure worldwide. Second, fMRI is noninvasive as both the magnetic fields and the RF (radiofrequency) power typically used in MRI have never been shown to cause detrimental health effects. Subjects have been known to volunteer hundreds of times over the course of years for fMRI studies, again with no ill effects.[1] Some consider the loud noise—greater than 90 dB (decibels)—to be potentially damaging long term with frequent and multiple repeated sessions, so this could be considered the most invasive aspect of fMRI. Earplugs mitigate acute effects however. Third, fMRI has higher spatial resolution and spatial certainty then PET, EEG, and MEG. No inverse problem of localization, central to EEG and MEG, exists. MRI has shown functionally informative voxel volumes as low as 0.5 mm × 0.5 mm × 0.5 mm. Fourth, fMRI is fast. Regarding speed of acquisition, the time between sequential imaging volumes can be as low as about 200 ms. The hemodynamic signal, on which fMRI is based, introduces a temporal and spatial smoothing to fMRI data; however, these hemodynamic limits are not prohibitive for a wide range of investigations and applications. Temporal delay differences on the order of 100 ms and functional activation foci at the columnar and layer levels have been mapped with fMRI. Fifth, fMRI has sufficient sensitivity to create functional maps on individuals in minutes—from a single run or even in seconds following a single period of activation. A typical scan session is about an hour, allowing implementation of either multiple task or stimuli manipulations or averaging of multiple identical runs to further increase signal detection.

Scientists who perform functional brain imaging are always searching for better methods to answer "where, when, and how much" regarding the assessment and mapping of neuronal activity. Functional MRI fills at least a spatial resolution niche that, serendipitously, matches well the systems-level organization of the brain that had previously been relatively unexplored. This spatial/temporal niche spatially is as fine as 1.0 mm and temporally covers 0.1 sec to hours or even days. Figure 1 shows where fMRI resides relative to other human brain

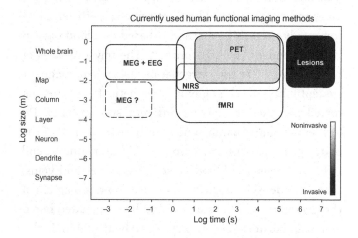

Figure 1 Spatial and temporal resolution and invasiveness of current human brain imaging methods: magnetoencephalography (MEG), electroencephalography (EEG), near-infrared spectroscopy (NIRS), positron emission tomography (PET), and functional MRI (fMRI). Recent work with MEG has suggested an ability for layer-level mapping—with some assumptions about cortical models.

imaging methods in terms of spatial resolution and temporal resolution.

This chapter discusses the basics of functional MRI, providing a brief overview of the basics of the functional contrast, some details of acquisition, and some aspects of the signal that have been found to be most interesting over the years.

Functional MRI uses magnetic resonance imaging to detect and map localized cerebral hemodynamic changes that occur at the location of increased brain activity. These localized hemodynamic changes include blood flow, volume, and oxygenation. The most sensitive, efficient, and commonly used fMRI method is based on blood oxygen level–dependent (BOLD) contrast. The physical basis of BOLD contrast is the oxygenation-dependent magnetic susceptibility of hemoglobin. Magnetic fields are distorted by materials having different susceptibility. Deoxyhemoglobin is paramagnetic relative to the rest of the body, thus setting up microscopic field distortions around red blood cells and blood vessels (capillaries and veins) that carry deoxyhemoglobin. Figure 2 shows a diagram of the magnetic field lines unaffected by oxygenated hemoglobin and distorted by deoxygenated hemoglobin. It also shows the hemodynamic changes that occur with brain activation. Because the precession frequency of protons is directly proportional to the magnetic field that they are experiencing in their microenvironment, these spatial

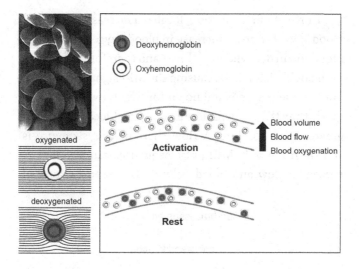

Figure 2   Red blood cells (top left). Shown here are the undistorted magnetic field lines around oxygenated hemoglobin and the distorted field lines around deoxygenated hemoglobin (bottom left), as well as the local hemodynamic changes that occur with brain activation including increased blood volume, flow, and oxygenation.

disturbances cause protons experiencing these field distortions to precess at slightly different frequencies, creating a signal cancellation as they become out of phase within an imaging voxel—leading to a slightly attenuated MRI signal that increases in attenuation the longer one waits to let the signal evolve. The waiting time is called the echo time or TE.

During brain activation, localized increases in arterial blood flow lead to an increase in blood oxygenation and a reduction in deoxyhemoglobin—and therefore reductions in the field distortions, causing the MRI signal to increase. The precise reason is unknown for why flow increases so much that the oxygenation in the blood locally becomes greater than baseline. A flow chart of this sequence is shown in figure 3. MRI pulse sequences can also be sensitized to flow and blood volume. However, because of

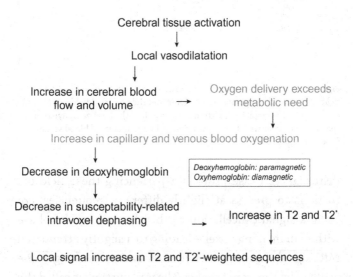

Figure 3 Flow chart of the relationship between neuronal activity and BOLD contrast. The two key components shown in lighter print are that oxygen delivery exceeds metabolic need and that hemoglobin changes susceptibility depending on its oxygenation.

their lower sensitivity, lower temporal resolution, more challenging implementation, and lower efficiency, these methods have not been embraced as readily as has BOLD contrast.

Functional MRI involves the use of an MRI scanner equipped with hardware and software allowing it to scan rapidly and repeatedly over time. The primary difference between MRI and fMRI is that with fMRI, an entire volume of data (typically thirty-five planes or "slices" covering the entire brain) is collected within about 2 sec using a pulse sequence called echo planar imaging (EPI). Multiple time series of brain volumes are typically collected over the course of about five minutes. Brain activation of durations from fractions of seconds to minutes is repeated during each time series scan, alternating with other brain activation tasks or stimuli (otherwise known as conditions) or periods of "rest" that usually involve fixating on a small cross at the center of blank screen. As already mentioned, activation leads to signal increases of a few percent in the activated region. Typical fMRI signal changes are shown in figure 4. Statistical analyses are performed on this time series signal to determine which regions demonstrated significant signal changes during brain activation.

Prior to the discovery of fMRI, several seminal findings helped lay the groundwork. Back in 1890, Roy and Sherrington first established that brain activity was

Functional MRI involves the use of an MRI scanner equipped with hardware and software allowing it to scan rapidly and repeatedly over time.

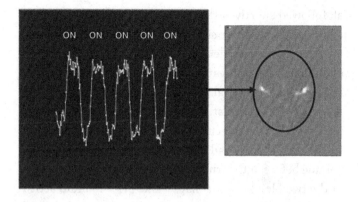

Figure 4 Typical BOLD contrast time series, collected at 1.5T of bilateral finger tapping. The image on the right is a simple subtraction of the average of the images during the on state (finger tapping) versus the average of the images during the nonactivated states (no finger tapping). The brain is outlined in black.

accompanied by localized hemodynamic changes.[2] In 1932, Linus Pauling first characterized the effect of oxygenation on blood susceptibility.[3] In 1982, Thulborn et al. determined that changes in blood oxygenation changed the transverse relation rate or T2 due to bulk susceptibility contrast.[4] In 1986, Fox and Raichle using PET demonstrated that the oxygen extraction fraction locally *decreased* with activation—implying that blood oxygenation locally *increases* with brain activation.[5] In 1989 and 1990, Ogawa et al. published a series of papers demonstrating blood oxygen level–dependent contrast in simulations, test tubes, and animals—coining the acronym "BOLD."[6] Turner et al. followed shortly with a dramatic demonstration of an oxygenation-sensitive signal decrease with apnea in cats.[7]

The first paper demonstrating MRI-based mapping of human brain activation was published in November 1991 by Belliveau et al.[8] However, this approach did not employ BOLD contrast. These results were based on the use of successive injections of an exogenous susceptibility contrast agent—gadolinium—to create maps of blood volume before and during visual stimulation. Subtraction of the two blood volume maps—one prior to brain activation and one during brain activation—depicted a localized increase in blood volume in the visual cortex during visual stimulation. This slightly invasive and relatively cumbersome technique became obsolete for brain activation mapping perhaps the moment it was published, given that the first experiments in blood oxygenation–based fMRI had already taken place. However, in and of itself this invasive blood volume mapping approach to simply map blood volume and the perfusion information that may be derived from it caught on clinically because it is a relatively quick and sensitive method for creating useful MRI maps of vascular patency that help diagnose susceptibility to and damage from a stroke.

The first experiments that demonstrated that MRI could be used—without any exogenous contrast agents—to map brain activation at relatively high spatial and temporal resolution were carried out in May 1991 by the research group at Massachusetts General Hospital (MGH) and during the summer of 1991 by the research group at the University of Minnesota. In September 1991, the group at the Medical College of Wisconsin observed their first positive results. I was a graduate student at the Medical College of Wisconsin and was carrying out those experiments with my fellow grad student, Eric Wong. Eric had been developing, designing, and constructing local gradient coils, RF (radiofrequency) coils, and pulse sequences for performing EPI. Once we saw the preliminary results coming out of MGH at the SMR meeting in San Francisco in August 1991, we were able to produce functional MRI maps within weeks. The first papers using blood oxygenation contrast to map brain activation were published in the early summer of 1992.[9] The term "fMRI" was coined in a paper published in 1993 on functional MRI processing strategies.[10]

Functional MRI was first performed on a variety of platforms: the MGH group used a retrofitted whole-body resonant advanced NMR (ANMR) gradient system at 1.5 Tesla (T). The Minnesota group was using a non-EPI multi-shot method at 4T. Our Milwaukee group was using a home-built head-only gradient coil and a home-written EPI pulse sequence on an otherwise standard General Electric 1.5T scanner. Gradient coil configurations used by our MCW group are shown in figure 5. Regarding the MGH and MCW results, the hardware used had been

1991–1992

1992–1999

Figure 5 Two versions (earlier version on top and later version on bottom) of the three-axis head gradient coil used for EPI by the Medical College of Wisconsin. These gradient coils were created by Eric Wong. The slew rate using 100 amp gradient amplifiers was 200 mT/m/ms, and gradient strength was 20 mT/m. The coil was made from sewer pipe, wire, and epoxy.

superseded quickly by the powerful commercially available gradient amplifiers, and for the Minnesota group, the multi-shot method has never gained popularity due to its relatively high time series instability and slower speeds. Multi-shot spiral acquisition was briefly popular as it was the most temporally stable multi-shot method at the time. However, it was still significantly slower than EPI. Multi-shot methods are making a comeback though, as higher resolutions that they bring are desired and better ways to correct the instabilities of multi-shot imaging are advancing. The now commercially available EPI approach still remains the most popular approach for fMRI today. Initially, none of the groups used sophisticated (or any!) statistics as the activation-induced signal change was quite easily detectable by eye.

Since the inception of fMRI, its use has grown in several ways that include an increased confidence in the meaning and neural correlates of the signal, increased sophistication of processing the data, increased number of users, and an ever-wider range of applications. Despite this success, fMRI has yet to make significant inroads with regard to day-to-day clinical implementation—mostly due to the variability of the time series signal as well as the variability in activation pattern across subjects and populations and individuals.

Resting State fMRI

In 1992 an explosive revolution in neuroimaging was precipitated by the discovery of fMRI. In 1995, a slow-to-ignite revolution within fMRI started with the discovery of "resting state" fMRI. While most fMRI applications have involved imposing specific brain activation tasks or stimuli on subjects and then observing their corresponding brain activation during time series collection of echo planar images, a growing area of fMRI has involved the collection of time series data while having the subject performing no task at all. Presently there are almost as many resting state fMRI papers per year as those involving imposed tasks. Resting state fMRI is based on the fundamental observation that the brain, when not engaged in any specific task, is never quiescent. Regions are spontaneously and transiently activated and are constantly interacting with other related or "connected" regions. These spontaneous activations and interactions are reflected in the temporal correlation of the time series signal from disparate yet connected (or at least coactivating) regions. In such a manner, "connected" regions of the brain can be identified purely from the resting state spontaneous activity.

Resting state fMRI began with the seminal observation by a graduate student at the Medical College of Wisconsin by the name of Bharat Biswal. He observed that the

Resting state fMRI is based on the fundamental observation that the brain, when not engaged in any specific task, is never quiescent.

fMRI time courses from a completely unengaged "resting" brain showed temporal correlation of the signal across related functional units[11] such as left and right motor cortex. In his study, subjects simply fixated on a small crosshair in the center of a blank screen. During this "resting state," a time course from the left motor cortex, appearing as typical fMRI time-series noise, was used as a "reference function." The correlation of the reference function with all the time courses from the entire brain were calculated and mapped, revealing that the other motor areas—supplementary motor cortex and the motor cortex in the other hemisphere—showed the highest correlation with this data-derived noise-like reference function. In other words, the resting brain was, in fact, not really resting, but showing enough synchronized spontaneous activity between functional units to influence the resting state signal to such a degree that the BOLD time series across these functionally connected regions showed a relatively high level of correlation even though no overt task was being carried out. Figure 6 shows, on the left, a map and time course derived from alternating twenty-second finger tapping on left and right hands. The signal from left and right motor cortex are therefore anticorrelated. On the right is a resting state map showing time series from the same regions that at first look like noise but exhibit a high level of correlation during "rest" as shown in the time

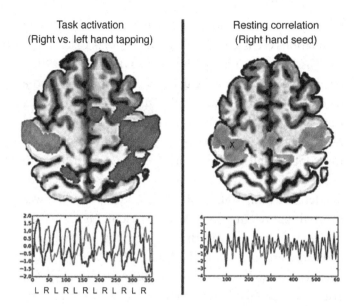

Figure 6 Activation (on left) vs. resting state correlation (on right). During the activation paradigm, the subject tapped their left fingers for 20 sec, then their right fingers for 20 sec, and repeated this sequence six times. The time course from the right and left motor cortex are shown below. The activation (dark blobs) is shown, illustrating left (light gray) and right (dark gray) activation. With resting state correlation analysis, the subject performed no task during the time course. However, signals from the right and left motor cortex, shown at the bottom, reveal high correlation even though they appear noise-like. The correlation map with a seed voxel in left motor cortex (indicated by an X) is shown, revealing both the left and right motor cortex. The primary frequency at which the correlation occurs is about 0.1 Hz.

series plots. The predominant frequency at which these correlations occur is in the range of 0.01 to 0.1 Hz.

The paper describing this study was published in 1995, but had no immediate influence on the direction of the field at the time it was published. The rate of publications demonstrating or using resting state fMRI remained relatively low for the next *ten* years—with no more than five papers published per year in resting state fMRI until about 2005. Finally, at this point, the rate of paper production in resting state fMRI began its steep climb. It was realized that resting state fMRI was useful for more than mapping connectivity in the primary motor, visual, and auditory cortex. The rapid advance after this quiescent period may also have been due to the practical reality that acquisition and processing methods had finally reached a state of sophistication such that detection of these resting state correlations was sufficiently robust and relatively straightforward. Also, back in 1995, the field of fMRI was still in its infancy. Researchers had barely begun in-depth activation-based studies, so were likely not ready to embrace simply looking at the noise during "rest."

Brain connectivity maps in the absence of task performance appear to match known brain circuitry at the systems level, including sensorimotor, visual, auditory, and language-processing networks. Among these networks, a relatively new and still somewhat mysterious "default mode" network was discovered. This brain network includes the posterior cingulate cortex (PCC) and medial prefrontal cortex (MPF). This finding dovetailed well with previous literature reports of a "default network" that been found to be most active when subjects are not engaged in any particular task. The precise functional role of this network remains unclear—as it has been associated with introspection, rumination, planning ahead, and self-referential cognition.

Many other networks have been found using resting state connectivity, allowing—with the complementary use of structural and diffusion tensor MRI—a comprehensive parcellation of the entire brain into "functionally connected" regions. These connectivity patterns derived from resting state fMRI have been shown to be repeatable across subjects and scans. With about ten minutes of averaging resting state data, stable connectivity maps emerge.

Once resting state fMRI gained in popularity, an increase in encouraging results quickly occurred. Many more resting state networks from "higher" cortical regions were found. New studies demonstrated that these networks are reflective of known cortical functional organization. Modulation in resting state correlations were demonstrated with cognitive interventions. Different clinical populations were shown to have different resting state correlation patterns or networks. Connectivity networks have been shown to correlate with behavioral characteristics such as fluid intelligence.[12] Applications of resting state fMRI to various brain diseases have been demonstrated, including studies of Alzheimer's disease,[13] schizophrenia,[14] epilepsy,[15] cocaine dependence,[16] and antidepressant effects.[17] It was found that patients with Alzheimer's disease showed decreased resting state activity in the posterior cingulate and hippocampus, suggesting disrupted connectivity between these two brain regions, consistent with the posterior cingulate hypometabolism commonly found in previous PET studies of early Alzheimer's disease. These studies that demonstrated the utility of resting state functional connectivity in the study of neurological and neuropsychiatric disorders and have ushered in a new era of fMRI aimed at clinical implementation.

Advanced approaches for analyzing resting state fMRI data, beyond the traditional seed-based and independent component analysis (ICA) methods, have been developed in recent years. Quantitative analysis of complex networks, based on graph theory, has been successfully exploited to study brain organization. Graph theory analyzes "graphs" consisting of nodes (e.g., anatomical regions of the brain) and edges (e.g., functional connectivity strength) connecting the nodes. It has been shown that brain systems exhibit topological features characteristic of complex networks, including "small-world" characteristics,[18] modular structures,[19] and highly connected hubs.[20] For example, a small-world network possesses both high clustering and short path lengths, resulting in efficient information transfer on both local and global scales. The human brain, like many other networks such as the internet social network and electrical power grids, has small-world properties representing a balance between integration and segregation between subunits. Network analysis methods have been demonstrated to be useful in identifying dynamic changes of brain networks associated with development,[21] aging,[22] and neuropsychiatric diseases.[23]

While the time-averaged resting state behavior converges to known stable networks, the resting state time series has also been shown to be dynamic,[24] with specific network correlation configurations emerging then disappearing in the span of seconds to hours. Studies have started to capture this variability and to attribute these variations to identify specific clinical populations—and importantly, to group individuals into specific clinical populations.[25] Methods for identifying this variability evolved, starting with wavelet analysis,[26] and then moving to sliding window pairwise correlation analysis between up to 500 predetermined segments, regions, and/or networks. The "pairwise correlation matrices" that are produced reveal the correlation of every region (or network) with every other region or network. This pairwise correlation matrix approach has emerged as a useful tool for comparing and displaying network activity at any moment in time.

The field of resting state fMRI is growing because the detected signals are robust and appear to be sensitive to a wide range of behavioral characteristics, brain states, and clinical symptoms. However, the precise neural mechanisms of "connectivity" and the underlying biologic role of resting state correlations remain a mystery in neuroscience. Another open question is related to the information contained in the time series fluctuations: What fraction of these fluctuations contain information about ongoing thought process, unconscious network communication, stable synchronous firing, and vigilance or arousal? Lastly, an ongoing challenge in resting state fMRI is that of more completely removing non-neuronal fluctuations from the resting state time series. Fluctuations from movement, cardiac pulsation, breathing, and hardware instabilities lead to false positives as well as missed correlations. A large effort is underway in the field to find better, more robust methods to remove such non-neuronal noise. Because the contrast to noise in resting state fMRI is nearly 1, and because the fluctuations overlap in space and frequency with many artifactual changes, the task of removing all artifact and noise is a daunting challenge.

Large, multicenter resting state fMRI databases have allowed researchers from around the world to apply different analysis approaches to mine this extremely rich resting state as well as activation-induced fMRI data. In this context, discovery science has value as the data contain information that may be missed by one avenue of investigation associated with each individual data set. Resting state functional connectivity is still in its infancy and is certain to grow in sophistication and use over the years—making inroads to aid in clinical diagnosis and prediction of treatment outcome as well as providing fundamental insights into the human brain and how it differs across populations and individuals.

fMRI Today

For researchers who study the brain and human behavior, the ability to quickly and easily obtain structural or functional correlates of a wide array of traits or symptoms makes functional brain imaging a powerful and attractive methodology. This ability has enabled researchers to draw inferences on how the healthy brain is organized on the systems level and has also been useful in understanding which brain regions or brain networks—at the systems scale—are not functioning properly in the context of disease or trauma.

As of the writing of this chapter (December 2018), the number of papers published in fMRI totals almost sixty thousand since 1991. Currently, thousands of papers are being published per year just in fMRI. Figure 7 shows the

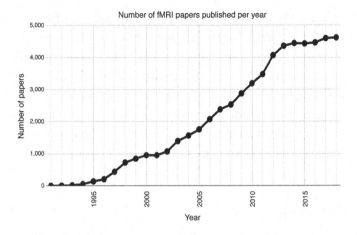

Figure 7 A literature search on the term "fMRI" (or "functional MRI") shows a steady upward trend with some tapering since 2013 in the number of papers published per year.

growth of the field as depicted by the number of papers published per year.

Of these publications, a large fraction includes results from group-averaged normal populations. A large and growing number of these studies are also group comparisons between clinical and normal control populations. Lastly, a small yet rapidly growing number of these publications concern methods that are moving in the direction of *individual subject* classification. It's relatively easy to draw inferences about whether populations show similar or different activation patterns. It's much more difficult to determine, using fMRI—or any neuroimaging method—if a single subject belongs to one population or another. For example, the field aims to someday place a person in the scanner, and after perhaps a resting state run and a series of diffusion tensor images and structural images, make an assessment with clinically useful probability (>80% likelihood for instance) that the subject has a psychological disorder or will best respond to a specific kind of therapy or drug. Single-subject classification is a primary application that will be necessary for fMRI to make a greater clinical impact.

In popular press, functional and anatomic brain images are ubiquitous. The dramatic and colorful images that are produced make for a compelling and informative visual display. However, they can lead to a false impression that all brains processes are now transparent to scientists and clinicians, and that if we can see where the activation occurs, we somehow understand what is going on in a deeper and more profound way. One way to think of the challenge of fMRI is with the analogous situation one would have if, when flying over a city at night, an attempt is made to determine the city activities in detail by simply observing where the lights are on and how they change. The information is extremely sparse, but with time, specific inferences can be drawn.

The study of the brain has permeated public awareness, but there has been something of a backlash in the

Think of the challenge of fMRI with the analogous situation one would have if, when flying over a city at night, an attempt is made to determine the city activities in detail by simply observing where the lights are on. The information is extremely sparse, but with time, specific inferences can be drawn.

scientific community in several areas. First, as mentioned, simply describing *where* an activity is occurring might be clinically useful, however, it does not imply that we understand the principle that determines why an area was activated or have any insight into any deeper organization principle. It's pure cartography. It also does not imply that we know what calculation—if we can even call a brain process that—is occurring in that area. Second, several popular articles as well as scientific papers have fallen into the trap of reverse inference. Functional brain imaging experiments allow the neuroscientist to infer something about the role of brain regions as they relate to cognitive functions. These data are closely tied to the specifics of the experiment. However, at times, brain activation maps have been used to infer the engagement of cognitive functions based on activation brain regions—or in other words, reverse inference. This reasoning is typically misleading because the same brain regions may subserve many different cognitive functions. In other words, just because, say, the amygdala shows more activation if a subject is viewing one presidential candidate rather than the other, it cannot be inferred what the subject feels about each candidate based on the degree of amygdala activation because there are just too many other variables. The amygdala is active for a wide range of cognitive states. Perhaps a way beyond this limit is by assessing the networked patterns of activation in more detail—likely resulting in more unique correlates to behavior. Such "decoding" approaches have grown in popularity due to their effectiveness in differentiating extremely subtle differences in stimulus or corresponding behavior. Progress has been made to educate both scientists and the public on the necessity to be careful in the interpretation of functional brain images and to not be too quickly swayed by the convincing aura of these compelling maps.

In clinical practice, anatomic MRI has been used as an effective brain imaging tool since about 1984. MRI scanners are in just about every hospital because their images are indispensable in identifying lesions, hemorrhage, swelling, tumors, and flow deficits in individual patients in a relatively quick and noninvasive manner. Functional MRI has made minimal clinical inroads. It currently is used clinically—in a highly limited manner—for presurgical mapping: identifying regions that are associated with primary motor, somatosensory, visual, auditory, or language function before surgical resection of a tumor or epileptic region. Functional MRI has much more potential, and perhaps as biomarkers are discovered, machine learning approaches to fMRI data analysis will augment a wider range of clinical applications.

THE SOURCES OF
FUNCTIONAL CONTRAST

The time course of several physiologic variables can be
measured and mapped with fMRI. Blood oxygen–level dependent (BOLD) contrast is a measure proportional to the
amount of deoxyhemoglobin in a voxel. Activation typically increases oxygenation locally, therefore decreasing
the amount of deoxyhemoglobin. MRI can also be sensitized to blood volume, blood flow and/or perfusion, and
can be further sensitized to hemodynamic changes at specific vessel sizes. Quantitative measures of changes in the
cerebral metabolic rate of oxygen ($CMRO_2$) with brain activation can also be derived. New methods have also shown
promise in mapping vessel radius in a voxel, mapping vascular territories, and mapping baseline blood oxygenation
and metabolism. Methods that are theoretically possible
yet still not convincingly demonstrated are direct mapping of neuronal current changes, mapping of neuronal

cell swelling with activation using diffusion-weighted imaging, and mapping of localized temperature changes. Just recently, a method that has claimed to measure elastic changes in the brain during activation was introduced at the 2017 International Society for Magnetic Resonance in Medicine Conference. In this chapter, the focus is on the established contrast mechanisms: blood volume, blood flow, blood oxygenation, and $CMRO_2$.

Blood Volume

In the late 1980s, the ability to obtain MR images in a fraction of a second emerged with the implementation of echo planar imaging (EPI). EPI required specialized hardware and was therefore not available on most clinical scanners. Vendors, however, were developing EPI as it held promise in the domain of cardiac imaging—allowing the creation of high temporal-resolution movies of the complete cardiac cycle. As a side benefit, the seeds for fMRI were sown as EPI would prove to lend itself well to fMRI for two reasons: its speed and its high temporal stability that allowed for the detection of small transient changes.

The ability to acquire a complete image in under 50 ms, and a complete volume in under 2 sec, allowed the tracking of MRI signal changes over time—a new dimension that, before 1990, was relatively unexplored. Rather than simply collecting one static image and comparing the image contrast over space, the rapid collection of identical images over time allowed assessment of dynamic changes in contrast.

An immediate application of this ability was the imaging of the transient effects of injected paramagnetic contrast agents. One could follow the MRI signal intensity as a bolus of a paramagnetic contrast agent such as gadolinium passed through the tissue of interest. Paramagnetic contrast agents in vessels concentrate the magnetic fields around their containing vessels, setting up microscopic field distortions, causing increased signal "dephasing" and therefore attenuating the MRI signal. As a bolus of gadolinium passes through the brain, the signal intensity is attenuated in proportion to the amount of gadolinium that is present in each voxel. Once the gadolinium washes out, the signal intensity increases to previous levels. The area under these signal attenuation curves are proportional to the relative blood volume.

In the late 1980s and early 1990s, for the first time, maps of the blood volume distribution in the brain using the gadolinium bolus injection method, with high diagnostic value, could be created.[1] In 1990, Belliveau and colleagues at Massachusetts General Hospital took this ability one step further and mapped blood volume *changes* during brain activation.[2]

Blood Oxygenation

It had been known for decades prior to the discovery of BOLD contrast that hemoglobin, the molecule that is essential for enabling red blood cells to carry oxygen to tissue, had unique properties. When it is bound to oxygen, it is diamagnetic, meaning it mildly repels magnetic fields. Diamagnetism is a property shared in almost equal amounts by all biologic tissue. When it releases oxygen, the hemoglobin becomes paramagnetic—having similar properties to gadolinium—and it concentrates and distorts the magnetic fields, causing signal attenuation. In fact, as deoxygenated red blood cells become less diamagnetic than plasma, veins become less diamagnetic than surrounding tissue. This difference in susceptibility at all these scales causes magnetic field distortions that lead to spins propagating at different frequencies and therefore dephasing.

In 1982, Thulborn et al. discovered that changes in blood oxygenation changed the transverse relaxation, or T2, of blood.[3] A short definition of T2 is that it is the rate at which the signal decays in the plane of detection or transverse plane after it is given energy or "excited" by an RF pulse. In tissue where the T2 is longer, the signal in a "T2-weighted" scan is brighter than in tissue where T2 is shorter. T2 is the transverse decay rate as measured using a "spin echo" pulse sequence. T2* is the transverse decay as measured using a "gradient echo" pulse sequence. $T2^*$ is almost always shorter than T2, as $T2^*$ is more sensitive to spin dephasing at all spatial scales.

A change in blood oxygen saturation therefore changes the MRI signal. Blood that is more oxygenated has longer T2 than blood that is less oxygenated. It was not until 1989 that this knowledge was used to image in vivo changes in blood oxygenation. Blood oxygen–dependent contrast, coined BOLD by Ogawa et al., emerged as a brain activation method of choice.[4] The first three papers showing human brain activation using BOLD were published within two weeks of each other in 1992.[5]

Interestingly, Ogawa predicted its utility for functional brain imaging in his earlier papers, hypothesizing that with brain activation the signal should change. However, he predicted a *decrease*, hypothesizing that during an increase in metabolic activity in the brain, more oxygen would be removed from the blood, thus causing a decrease in T2 and a signal decrease.[6] A few years earlier, however, a paper based on positron emission tomography (PET) suggested that an increase in brain activation is accompanied by a large increase in blood flow to the active area— overcompensating for any increase in oxidative metabolic rate. Therefore, with an increase in brain activation, the metabolic rate increases, but the large increase in oxygenated blood flow to the area causes the overall oxygenation to increase despite the increase in the oxidative metabolic rate. This increase in oxygenation causes an increase in T2 and T2* relaxation times, and thereby an increase in signal in T2- and T2*-weighted sequences. In fact, this is precisely what had been seen with fMRI—an activation-induced signal increase of about 1%–5% at 1.5T when using gradient echo imaging, suggesting such an increase in blood oxygenation.

T2 and T2* decay curves are shown in figure 8. Most BOLD contrast imaging today uses T2* contrast because it is more sensitive than T2 contrast to blood oxygenation changes by at least a factor of 2 to 4. This is mostly because T2* changes pick up susceptibility-induced inhomogeneities from small to large perturbers. Spin echo sequences

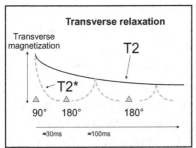 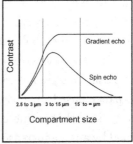

Figure 8 The left panel shows T2 and T2* decay curves following an excitation pulse. T2* decay is more rapid than T2. A T2*-weighted image is formed using gradient refocusing during the T2* decay—here shown at about 3 ms. A T2-weighted image is formed using a 180-degree RF pulse to refocus the magnetization. The right panel shows the relative spin echo and gradient echo sensitivity to susceptibility compartment size.

are more sensitive to small perturbers on the order of scale of a red blood cell. Spin echo susceptibility contrast relies on protons diffusing through the field perturbations to have an effect on the signal. In the short time that a spin is allowed to diffuse during imaging, it will more likely diffuse through small sharp perturbations from red blood cells and capillaries. Since there is more area covered by the large perturbations (i.e., larger veins), the gradient echo signal will dephase more protons and result in a larger signal change.

The best echo time to use in an imaging experiment intended to optimally detect changes in $T2^*$ directly comes from the fact that the signal decay is an exponential. During rest, a single exponential with decay rate $T2^*$ describes the transverse magnetization. During activation, the $T2^*$ decreases slightly, shifting the exponential decay. If one were to compare the signal between these two exponentials, one would find that the percent change between these continues to increase with TE, but the difference between the exponentials peaks at approximately the TE equal to the resting state $T2^*$. Signal contrast is fundamentally the signal difference and not the percent change; therefore, in all fMRI studies the TE used is approximately equal to the $T2^*$ (or T2 if spin echo acquisition is used). This depiction of the optimal TE to use is shown in figure 9.

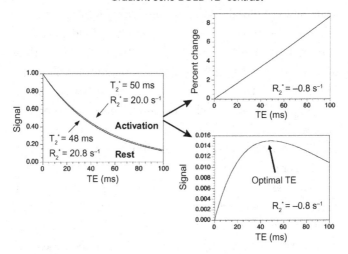

Figure 9   The graph on the left shows the transverse signal decay as a function of TE for rest and activation, assuming a baseline T2* of 50 ms and a change in T2* of about 2 ms. As a side note R2* is simply equal to 1/T2*. If one calculates the percent signal change as a function of TE, it increases linearly, as shown in the top right graph. If one calculates the difference between the two exponential decays, it peaks at about TE = T2* = 50 ms. This is the reason why, in fMRI, the TE chosen is equal to T2*, as it optimizes the signal difference that determines the functional contrast.

Blood Perfusion

BOLD was not the only functional contrast to emerge in the early 1990s. MRI-based perfusion imaging, also known as arterial spin labeling (ASL), was more developed than discovered. It is a method that allows the creation of noninvasive, quantitative baseline perfusion maps as well as maps of activation-induced perfusion changes. In fact, Kwong et al. in their first fMRI paper published in the *Proceedings of the National Academy of Sciences (PNAS)* in 1992 also demonstrated that a T1- (or longitudinal relaxation–) weighted scan could detect subtle T1 changes that occur with activation-induced localized perfusion changes.

T1 is a measure of the rate in which the magnetization returns to equilibrium after it is excited by an RF pulse. It is always longer than T2, as T2 is simply a measure of how quickly the signal is dephased in the transverse plane. The net magnetization may still not be fully returned to equilibrium even though it is completely dephased. If perfusing spins from outside the imaging plane enter into the plane after excitation, they will add to the longitudinal magnetization given that they are fully at equilibrium already, thus causing the overall T1 in the plane experiencing inflowing or perfusing blood to appear more fully relaxed, with a shorter T1 (more rapid return to equilibrium). If perfusion increases, T1 shortens and the signal in a T1-weighted scan gets brighter. In this manner perfusion rates can be measured using T1-weighted scans, as was done in the Kwong paper mentioned earlier.

ASL-based perfusion mapping methods do something slightly different. They "label" the inflowing magnetization and observe the effects of that label on the imaging plane. They are similar, in principle, to tracer methods applied in other modalities such as positron emission tomography (PET) and single photon emission computed tomography (SPECT) in that they involve "tagging" or "labeling" inflowing blood, and then imaging the effects of the tagged blood as it moves into the imaging plane. Instead of tagging blood with an injected contrast agent, the "tag" is provided by a spatially selective radio frequency pulse to inflowing blood, altering its magnetization. The RF tagging pulse is usually a 180 degree pulse that "inverts" the magnetization while it is outside the imaging plane. Once the blood flows into the plane of interest, the tagged blood changes the magnetization of the tissue where it is perfusing and thus interacting with the protons in the plane of interest. Perfusion images are created by subtracting images created without the tag from images with the tag. These images highlight only the signal that changed with the tag. The signal changes in each voxel of the images are proportional to perfusion.

The advantage of perfusion imaging is that it is a direct and potentially quantitative measure of activation-induced flow changes. It also provides clinically useful baseline perfusion maps—easily differentiating gray and white matter which has a difference in perfusion of a factor of 2 to 4. The disadvantages are that it is less sensitive to BOLD contrast by a factor of 4. However, it has been put forward that it is optimal for long-duration brain activation as the time series signal does not have slow drifts in it as the BOLD signal typically has. Perfusion imaging also has intrinsically lower temporal resolution because the added necessity for the "inversion" 180 degree pulse forces the repetition time (TR) to be at least 1 sec. Lastly, brain coverage using ASL methods is limited to a single slab at a time that does not fully cover the brain, so whole-brain imaging is much more cumbersome.

Blood Volume Imaging without Contrast Agents

In the early 2000s, Hanzhang Lu developed vascular space occupancy (VASO), a method for imaging blood volume changes without the need for exogenous contrast agents. The pulse sequence made use of the understanding that blood T1 (or "longitudinal relaxation"—the amount of time it takes for the net magnetization to relax back to equilibrium—typically much larger than T2 or T2*) is different from that of brain tissue. If a 180 degree pulse (a complete inversion of inflowing spins) is applied, then, after a specific amount of time depending on the T1 of the tissue or blood, the signal will start out negative and then pass through what's known as a "null point" where it is invisible. After passing through the null point, the signal will become positive until it is fully recovered. This null-point time is different between blood and tissue, and therefore if one waits until the blood passes through the null point, there will be a signal void (from the invisible blood) that is proportional to the amount of blood in the voxel. With brain activation, the blood volume increases, which means the relative volume of the signal void will increase, thus lowering the overall signal in that voxel. Many have shown that VASO is very selective to capillary- and small vessel-specific effects—as these show the clearest blood volume changes—and therefore is proving itself useful for extremely high-resolution fMRI where layer-or column-dependent activity is desired.

Cerebral Metabolic Rate of Oxygen ($CMRO_2$)

In the late 1990s and early 2000s, advances were made allowing the mapping of activation-induced changes in the cerebral metabolic rate of oxygen ($CMRO_2$). The basis for such measurement starts with the understanding that blood oxygenation is sensitive to opposing influences: flow increases (increase in flow leads to localized increases in oxygenation and therefore causes an increase in signal) and metabolic rate changes (increase in metabolic rate without a flow increase would decrease oxygenation and therefore decreases signal). With brain activation, increases in localized flow outweigh metabolic rate changes such that the overall oxygenation increases and therefore signal change is positive. Normalization using a hypercapnia has evolved into a method for directly mapping changes in $CMRO_2$. The basic idea is that when a subject is at rest yet undergoing a hypercapnic stress (5% CO_2), the cerebral flow increases without an accompanying increase in activated-induced $CMRO_2$ changes and therefore less oxygen is extracted from the blood stream than during brain activation. The ratio of BOLD signal changes to flow changes with a hypercapnic stress would be smaller than the ratio during brain activation, because with activation the increase in $CMRO_2$ removes some of the oxygen from the blood, thus blunting the activation-induced BOLD increase relative to hypercapnic changes. By comparing the ratio of the (simultaneously measured) perfusion and BOLD signal changes during hypercapnia and during brain activation, $CMRO_2$ changes with brain activation can therefore be derived.

The mapping of baseline $CMRO_2$ is more difficult as assumptions about blood volume in each voxel need to be made. So far, such techniques for mapping baseline cerebral oxidative metabolic rate have not been fully developed. Progress is being made toward this goal, however; with the advancement of better calibration techniques and fewer assumptions about how venous blood volume varies with each voxel, more precise assessments of baseline $CMRO_2$ can be made.

Since $CMRO_2$ mapping was developed in the late 1990s, it has not caught on as an fMRI method. While quantitative information on brain oxidative metabolism potentially could be useful, the method is cumbersome, involving CO_2 breathing, division of two noisy measures, and a specialized pulse sequence.

In the first section of this chapter, we provided an overview of fMRI contrast mechanisms. Now we will delve more into the empirical characteristics of fMRI contrast: its specificity, latency, magnitude, and linearity. These characteristics define the limits and potential of fMRI—the very limits of what we can do with the signal. Understanding what influences these limits is essential to a deeper appreciation of all aspects of fMRI and for effectively designing, carrying out, processing, and interpreting fMRI experiments.

Hemodynamic Specificity

The goal in all hemodynamic measures of brain activation is enhanced sensitivity to smaller vessels, which are more proximal to regions of activation. Larger vessels are more upstream (arteries) or downstream (veins) and may lead to a spatial misrepresentation of the true area of activation. They may also lead to misinterpretation of the magnitude of the signal change for a given oxygenation change, as the magnitude of the fMRI signal change is proportional to the venous blood volume in each voxel. Figure 10 depicts graphically this concept by showing how each voxel contains within it a different proportion of arteries, arterioles, capillaries, venules, and veins, and, accordingly, a different blood volume from each vessel. This will result in a large difference of fractional signal change—up to an order of magnitude difference—for a given blood oxygenation change. This underlying variation in voxelwise vasculature distorts the activation and causes spatially varying delays as different parts of the vasculature become oxygenated at different times as blood flows downstream.

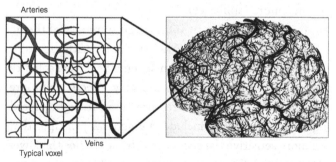

Each voxel contains a different fraction of arteries, arterioles, capillaries, venules, and veins.

Arteries

Veins

Typical voxel

Figure 10 Depiction of the sampling of the vasculature in an echo planar image with voxel dimensions on the order of 1–3 mm^3. For a given oxygenation change, each voxel will show not only a different signal change magnitude, but likely also a different hemodynamic latency.

While the spatial certainty concern has been raised over the years, it has not been a major obstacle for the success of fMRI as, typically, blobs of activation on the order of 1 cm, obviate the need for capillary-level precision. Recently this concern has been raised as high field imaging (and the sensitivity it brings) has allowed extremely high-resolution fMRI to be carried out such that noncapillary upstream or downstream signal changes would severely skew the spatial location of activation. If one is trying to delineate activity between the surface or layer activity and the deeper, layer-dependent activity of a segment of cortex, a spatial precision on the order of <1 mm is desired, making noncapillary influences extremely problematic. In particular for layers, there is always a blood volume gradient that goes from the surface (pial vessels) to the deeper regions (more capillaries).

It is useful to give an abbreviated summary of the hemodynamic specificity of fMRI methods. With regard to choosing the right acquisition scheme or pulse sequence for optimizing sensitivity to capillary effects, there are other factors in play that require some thought regarding trade-offs. These include the general principle that the more sensitive the sequence is to capillaries the lower sensitivity it will have overall since capillaries fill at most about 2%–4% of a voxel while larger vessels may fill a voxel 100%. Also, several capillary-sensitive techniques such as VASO and ASL are limited in spatial coverage (they can only cover a few slices per acquisition) and time (the added "inversion" pulse adds to the overall time per image, thus lengthening the TR or time between each image acquisition).

Regarding BOLD contrast-dependent functional imaging, spin echo sequences are more sensitive to small susceptibility compartments (capillaries and red blood cells) and gradient echo sequences are sensitive to susceptibility compartments of all sizes. A small amount of diffusion weighting, or "velocity nulling," reduces the intravascular signal therefore reducing, but not eliminating, large vessel effects in gradient echo fMRI, but eliminating all large vessel effects in spin echo fMRI. At very high magnetic fields such as at 7 Tesla, blood $T2^*$ is very short—much shorter than the $T2^*$ of tissue. Therefore, even without diffusion weighting, intravascular signal is extremely small, so that spin echo sequences at high field are sensitive to capillaries without the need for additional measures to minimize intravascular signal. In general, with an increase in field strength, BOLD contrast increases linearly to slightly super-linearly and overall image signal to noise ratio increases linearly; therefore, higher field is desired for fMRI based on BOLD contrast—especially at high resolution.

Perfusion imaging is generally accepted to be more specific to brain activation-induced hemodynamic changes in capillaries, but perfusion methods also can suffer from intravascular signal contamination from larger inflowing arteries and arterioles. As mentioned, it has also been shown that VASO methods are highly selective to capillary effects—allowing precise localization to small regions of brain activation, demonstrating layer-dependent activity with sensory-motor tasks.

In almost all fMRI studies that have been performed, the need for sensitivity has consistently outweighed the need for specificity greater than 1 cm. With studies performed at high field and high-resolution fMRI, specificity carries more importance, and therefore there has been a renewed interest in further developing sequences that are more specific to capillary effects that are most proximal to true neuronal activation, even if sensitivity is sacrificed somewhat.

The Hemodynamic Transfer Function

The hemodynamic transfer function is referred to here as the combined effect on the fMRI signal change shape, latency, and magnitude by the neuronal-vascular coupling, blood volume, blood flow, blood oxygenation, hematocrit, and vascular geometry, among other variables. A goal of fMRI method development is to completely characterize this transfer function as it varies across subject populations, individual subjects, regions in the brain, and even voxels. A more ambitious goal is to not only character-ize this variability, but to also develop robust calibration methods such that studies may "look through" this vari-ability to derive more specific information from BOLD contrast.

Characterizing the transfer function is challenging be-cause the variables previously listed all can contribute and vary among subject, brain region, and voxel. The very idea of a transfer function also assumes a linear system, which to a first approximation characterizes the hemodynamic response. However, it tends to behave nonlinearly, provid-ing more contrast than expected with very brief (<3 sec) or very weak neuronal activation.

The goal is to allow fMRI to make more precise inferences about underlying neuronal activation loca-tion, magnitude, and timing. The ultimate limits of fMRI depend on being able to make maps of these variables—allowing spatial normalization and more precise infer-ences about neuronal activity. This is particularly relevant when attempting to compare subject populations or in-dividuals where these hemodynamic effects might vary. For instance, a medication may vary fundamental mecha-nisms of neurovascular coupling in a patient, thus render-ing any interpretation of differences in BOLD contrast between the patient and a nonmedicated normal volun-teer problematic. The issue of characterizing the hemody-namic transfer function is also relevant when attempting to infer such things as precise timing or causality between regions (i.e., how one region acts on or is acted on by another) where the hemodynamics may vary from voxel to voxel.

After the onset of activation, or rather, after the neuronal firing rate has passed an integrated temporal-spatial threshold, either direct neuronal, metabolic, or neurotransmitter-mediated signals reach arteriole sphincters, causing dilatation. The time for this initial process to occur is likely to be less than 100 ms. After vessel dilatation, the blood flow increases by 10%–200%. The time for blood to travel from arterial sphincters through the capillary bed to pial veins can be up to about 2 to 3 sec. This transit time determines how rapidly the blood oxygenation saturation increases in each part of the vascular tree. Depending on which part of the vascular tree is predominantly captured by each voxel, the latency, shape, and magnitude of the hemodynamic response might vary significantly. A typical hemodynamic response function is shown in figure 11. This function is typically modeled as a Gamma Variate function. This transfer function, in humans, is noted for a peak occurring about 5 sec after the onset of stimulation, and a post undershoot that follows activation.

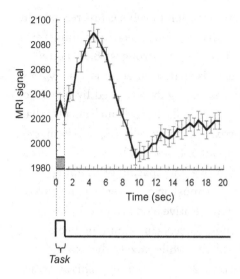

Figure 11   This is a depiction of a hemodynamic transfer function, otherwise known as the impulse response function. For most task durations >3 sec, the hemodynamic response behaves in a linear manner and therefore can be described fully by this function. To predict fMRI time series, this function is convolved with the neuronal input function.

Location Specificity

In resting state, hemoglobin oxygen saturation is about 95% in arteries and 60% in veins. The increase in hemoglobin saturation with activation is largest in veins, changing from about 60% to 90%. Capillary blood oxygenation changes from about 80% to 90% saturation. Arterial blood, already saturated, shows virtually no oxygenation change.

This large change in saturation in veins is one reason why the strongest BOLD effect is usually seen in draining veins.

The second reason why the strongest BOLD effect is seen in draining veins is that, as mentioned, activation-induced BOLD contrast is highly weighted by blood volume in each voxel. Since capillaries are much smaller than a typical imaging voxel, most voxels, regardless of size, will likely contain about 2%–4% capillary blood volume. In contrast, since the size and spacing of draining veins is on the same scale as most imaging voxels, it is likely that veins dominate the relative blood volume in any voxel that they pass through. Voxels containing pial veins can have 100% blood volume while voxels that contain no pial veins may have only 2% blood volume. This stratification in blood volume distribution, illustrated in figure 10, strongly determines the magnitude of the BOLD signal.

Different fMRI pulse sequences can give different locations of activation due their different sensitivities to specific vasculature. The ASL-based perfusion change map is sensitive primarily to capillary perfusion changes, while the BOLD contrast activation map is weighted mostly by veins near the activated region. For high-resolution studies, pulse sequences such as VASO (sensitive to blood volume changes predominantly in capillaries) and spin echo sequences (sensitive to small compartments including capillaries) at high field increasingly are being used to detect very small regions of activation.

In spite of the hemodynamic limitations of these methods, detailed activation at the cortical column level[7] and the laminar level[8] have been reported using BOLD contrast.

Latency

One of the first observations made regarding fMRI signal changes is that the BOLD signal takes about 2 to 3 sec to begin to deviate from baseline following the onset of activation. Since the BOLD signal is highly weighted toward venous oxygenation changes, with a flow increase, the time for venous oxygenation to begin to increase will be about the time that it takes blood to travel from arteries to capillaries and draining veins—about 2 to 3 sec. The hemodynamic "impulse response" function, as described, has been effectively used to characterize much of the BOLD signal change dynamics. The use of this function to predict the hemodynamic effect from all stimuli timings assumes the system is linear, which holds for most time scales but appears to deviate with stimuli durations shorter than about 3 sec. Regardless of this nonlinearity, observed hemodynamic response to any neuronal activation can be predicted with a reasonable degree of accuracy by convolving the expected neuronal activity time course with the BOLD "impulse response" function.

Early on, it was observed that BOLD latencies varied up to 4 sec in the brain. The variation was not as much between regions but more distributed across voxels. These latencies were also shown to correlate with the underlying vascular structure. The earliest onset of the signal change appeared to be in large-vein-free gray matter and the latest onset appeared to occur in the largest draining veins. Similar latency dispersions in the motor cortex and other areas have been observed.

As a side note, the overall latency distribution of resting state fluctuations has been shown to vary between gray and white matter and with vasculature. In fact, this method of observing latency with resting state has been implemented clinically to assess regions of stroke or compromised vasculature. Areas that are compromised have more delayed effects and therefore longer latencies in their baseline fluctuations.

Methods for working around these spreads of latencies in the context of fMRI to delineate cascaded brain activity with cognitive tasks have included forcing specific cognitive components of tasks to be long (on the order of several seconds) so that they may be differentiated, or by simply looking at the *change* in the latency with specific and task timing changes. In the latter studies, the absolute latencies are not mapped but, rather, the spatially specific changes in latencies are determined as they correlate with the task timing modulation. Using the latter approach of looking at changes, either in latency or width, task timing modulations of as low as 50 ms have been discerned.[9]

Regarding the hemodynamic response itself, it seems to be quite sensitive to transient neuronal activity. Stimulus durations as low as 16 ms (limit of the refresh rate of the stimulus apparatus) have been shown to elicit a response.

Magnitude

The magnitude of the fMRI signal change is influenced by several non-neuronal variables that may vary across subjects as well as across voxels in each subject's brain. Within subjects, it varies from voxel to voxel, thus influencing the voxel-averaged magnitude across activated regions as well as the activation pattern within activated regions. A complete and direct correlation between neuronal activity and fMRI signal change magnitude, in a single experiment, will remain nearly impossible until all the variables can be characterized and/or calibrated on a voxel-wise basis. Because of these physiologic variables, brain activation maps will typically show a range of BOLD signal change magnitude from 1%–10% for any given stimulus. Typically, large vessel effects show much higher fractional signal changes, so a crude method to remove large vessel effects is to simply establish the threshold based on an upper bound of percent change. Regarding the interpretation of the magnitude of the fMRI signal change, the picture is not that bleak. Progress has been made in successfully characterizing the magnitude of the fMRI signal change *as it varies* with the degree of neuronal activity variations. Like modulating timing to see differences in latency with a timing variation, the signal change magnitude is commonly modulated in a systematic manner so that the influence of intrinsic hemodynamic weighting is minimized. Such experiments are commonly referred to as parametric designs and are quite powerful as any non-neuronal influences, including vascular structure as mentioned, do not change with neuronal activation and therefore can be subtracted.

Many studies have been published that have increased our confidence that BOLD magnitude is indeed able to be altered by a systematic modulation in the amount of neuronal activity. Inferred brain activation modulations—visual cortex modulation by altering the flicker rate or contrast of the display; motor activation modulation by altering finger tapping rate; and auditory cortex modulation by altering syllable rate—have resulted in clear monotonic BOLD signal change correlations. Parametric experimental designs represented a significant advance in the way fMRI experiments were performed, enabling more precise inferences about the BOLD signal change with task modulation.

The magnitude of the fMRI signal change is influenced by several non-neuronal variables that may vary across subjects as well as across voxels in each subject's brain.

In the early 2000s a seminal paper by Logothetis et al.[10] demonstrated a clear relationship between electrophysiological measures and BOLD contrast. Simultaneous collection of multi-unit-recording array data and BOLD-based fMRI data was carried out in primate-visual-cortex visual stimulation with a flashing checkerboard of varying contrast. The study demonstrated an approximately linear relationship between local field potential power and BOLD contrast—within a range of stimulus contrasts. However, the relationship became highly nonlinear at low levels of stimulation, where the BOLD signal overestimated the amount of neuronal activation. The linearity of BOLD responses can be considered in different ways, described in the next section.

Linearity

Here we delve a bit more into the BOLD signal as it behaves as a function of task timing and intensity. As mentioned earlier, it has been found that with very brief stimulus durations, the BOLD response shows a larger signal change magnitude than expected from a linear system. This greater than expected BOLD signal change is specific to stimuli durations below 3 to 4 sec. Figure 12 shows this greater than linear response behavior. Reasons for nonlinearities in the event-related response may be neuronal,

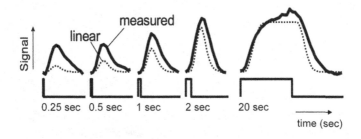

Figure 12 The hemodynamic response is greater than a linear system at durations below about 3 sec. Here is a depiction of actual data (solid line) and simulated data (dashed line) of the response to increasing stimulus durations (depicted by boxcars at the bottom).

hemodynamic, and/or metabolic in nature. For instance, it has been shown in electrophysiological measures that the neuronal response shows a large transient at the onset of activation. Nonlinearities in the response can also arise from mismatches in the timings of flow, volume, or metabolism changes. It has also been found that BOLD contrast is still detectable with an on/off oscillation of up to 0.75 Hz—also well above the threshold detectable frequency from a linear model of the hemodynamic response.[11]

BOLD contrast is highly sensitive to the interplay of blood flow, blood volume, and oxidative metabolic rate. If any of these variables has a rate of change that is different than the others, nonlinearities in the relationship between neuronal activity and BOLD may be manifest.

HARDWARE AND ACQUISITION

An extensive review of the physics and engineering underlying the acquisition of the rawest signal coming off the scanner and the creation of MR images is beyond the scope of this book. This chapter offers instead a basic, practical understanding of MRI hardware and acquisition suitable for anyone working with or interested in fMRI. So much of the signal as well as the artifact is tightly tied to the details and subtleties of how the scanner itself works and how the data were acquired. This chapter provides information and guidance to encourage readers to better design and carry out experiments and to have more useful conversations with individuals engaged in the fMRI acquisition process.

First, it's important to put almost all of fMRI technology in perspective. MRI technology has almost completely been driven by the powerful, clinic-based MRI market—and the same is true of fMRI, which uses clinical MRI systems with clinically-focused pulse sequences. However, MRI systems were not designed with fMRI in mind, which means that while the field of fMRI has benefited from this relationship, in many ways it is also limited by it.

MRI technology has been developed by academic centers in close contact with the industry leaders in MRI (principally Siemens, GE, Philips), then incorporated into the clinical scanners and distributed for clinical use. However, innovation for most fMRI technology has taken a slightly different route. Typically, fMRI R&D has been carried out by academic labs with engineers or physicists who have the skills and interest to modify MRI clinical pulse sequences, image reconstruction methods, and hardware such as RF coils and gradient coils. Since the current market for fMRI is only a fraction of the entire MRI market, research-directed innovations that do not spill off from clinical market-focused innovation tend to come from academia. These academic-driven innovations typically translate into a widely distributed product *only* if they are seen to benefit a more clinically relevant application. However, many if not most fMRI innovations in acquisition are the by-product of collaborating researchers, waiting to be converted into something that a wider range of people can use. Functional MRI as a field will advance significantly if

Functional MRI as a field will advance significantly if vendors invest more time and money in developing better systems optimized specifically for fMRI.

vendors invest more time and money in developing better systems optimized specifically for fMRI.

In the sections that follow, this chapter describes the practical nuts and bolts of MRI technology: The primary magnetic field (B_0), Radiofrequency (RF) coils, shim coils, gradient coils, and pulse sequences. An outstanding primer on all these variables also can be found in the self-published book *All You Really Need to Know about MRI Physics*.[1]

The Primary Magnetic Field: B_0

MRI is based on the fact that specific elements (hydrogen, deuterium, lithium, carbon, nitrogen, fluorine, sodium, phosphorus, and potassium) have a magnetic moment, meaning that when these elements are placed in a magnetic field (B_0), they have a distribution of energy states that can be modulated by an oscillating magnetic field (B_1), produced by an RF coil. Then, while in their "excited" state, these elements are measured with typically the same RF coil, now acting as an antenna. Hydrogen in water not only makes up a large fraction of the body mass but also has an extremely large magnetic moment. Therefore, almost every MRI produced is of water.

Each of these unique elements also have what is known as a gyromagnetic ratio, which determines the rate at which each element precesses or oscillates when experiencing the primary magnetic field, B_0. Their precession frequency is equal to the gyromagnetic ratio × B_0, known as the Larmor frequency. To give a point of reference, hydrogen precesses at about 42 MHz in a 1 Tesla field. The B_1 field pulse frequency pulse has to be in "resonance" or oscillating at the same frequency that provides energy at the "resonance" frequency of the element. This frequency is serendipitously within the completely-safe radio-frequency range, hence the reason for the term for the coil that generates the oscillating field.

The amount of signal we can detect is directly proportional to the magnetic moment. It is therefore a fundamental principle in MRI that sensitivity increases with field strength, or B_0, which increases the magnetic moment as it increases. Because sensitivity is a highly desired commodity in all MRI applications, and because high field strength is achievable by relatively straightforward technological advancements (i.e., more wire windings), field strength for human scanners has steadily increased since the beginning of human MRI. It should be noted that increasing field is not really that straightforward: superconducting wire technology is pushed to its very limits in the context of MRI scanner manufacturing. The wires need to be as small as possible yet able to carry the massive currents required for MRI scanners. Coolant technology needs to be robust and compact. All these factors drive up the price of high field scanners. While the current cost of high field scanners scales linearly with field strength—at approximately \$1 million cost per Tesla, and a bit higher at the extreme high end and a bit lower at 1T and below—the ability to extract novel contrast more efficiently and accurately and at higher resolution increases perhaps superlinearly with sensitivity, thus maintaining the drive to high fields.

Functional contrast to noise using BOLD also scales at least linearly with field strength. The type of anatomic MRI contrast possible also becomes qualitatively different at high field. For example, phase contrast at 7T is profoundly more sensitive to susceptibility contrast. With the advantages of high field, the researcher can also trade resolution or speed, or both, with the new abundance of sensitivity. Signal to noise scales linearly with voxel volume, so the practical lower voxel-volume limit should become smaller in proportion to field strength. At 3T, about 2 mm^3 voxel volumes are practically possible for fMRI; however, at 7T, 1.5 mm^3 is practically possible with fMRI.

The highest field strength for human MRI had been 11.7T at the National Institutes of Health. However, that scanner "quenched." When a magnet quenches, the supercooled superconducting wire starts to heat, thus increasing its resistance. With increased resistance, more heat is generated, thus an irreversible cascading increase in temperature commences, and an explosive "boiling off" of the liquid cryogens takes place. This process can damage the primary magnet and of course results in the loss of all of the extremely expensive liquid helium (a coolant), becoming quite costly. The 11.7T scanner was, in fact, damaged, and is now being repaired, so the highest field for now is 10.5T. The most common high field strength human scanner is 7T, with about sixty such scanners being used worldwide.

The achievement of extremely high resolution functional MRI at 7T has required innovation in several areas, given that four major challenges are manifest with higher field strength. First and foremost, transverse relaxation (T2* and T2) become shorter with high field, allowing less time for data acquisition after the excitation RF pulse before the signal decays into the noise. Second, magnetic field inhomogeneities, manifesting as signal dropout regions, are much more significant and are hard to remove. Third, RF excitation uniformity throughout the brain becomes worse. Without correction, RF flip angles vary from region to region—adding uncertainty to the structural and even functional contrast. Fourth—less of a problem for fMRI as it is for more RF intensive sequences—RF power deposition increases with increases in B_1. As mentioned, at higher fields, higher RF frequencies are needed to "excite" the protons—a first step in creating MR images. With higher frequency, increased RF power is necessary, resulting in potential tissue heating for high RF duty-cycle sequences, thus limiting the RF duty cycle in certain sequences for high-resolution anatomic imaging. Commercial scanners have an upper limit of RF power allowed, keeping the "specific absorption rate" (SAR) well below the level that would lead to tissue heating.

Many solutions or at least partial solutions to these challenges have been implemented. More rapid acquisition methods have helped to compensate for the fact that the signal dies away more rapidly at high field (short $T2^*$ and T2). To compensate for signal dropout, shimming hardware and techniques have generally improved and smaller voxels generally reduce any signal dropout. RF uniformity issues have been addressed by calibration scans in addition to new technology that allows adjustment of RF power from each coil "element" enabling "smoothing out" of the power distribution in the brain—also known as "RF shimming."

The achievement of extremely high-resolution fMRI at high fields has also introduced new challenges regarding how to average and compare fMRI data. Typical practice in the past has involved spatial smoothing, spatial normalization, and averaging of functional activation maps of multiple subjects. Performing spatial smoothing on high resolution removes the benefits of collecting at high resolution in the first place. Also, spatial variability of fine structures increases with increased resolution. It's much easier to spatially average large swaths of brain across subjects than to average, for instance, cortical columns. Large swaths of brain activation are similarly distributed across subjects, but finer details of activation have much more variability, defying spatial averaging. Rather than collapsing and comparing data spatially, it is likely possible to collapse and compare in other data dimensions or perhaps locally along columns or layers, thus keeping the detailed high-spatial-resolution information. Such innovations have yet to be fully realized. High resolution also represents a paradigm shift in that the research focus is on individual results with high resolution rather than spatially averaged group results coming from low-resolution scans.

Radiofrequency Coils

Radiofrequency (RF) coils provide the oscillating field that "excites" the protons that are imaged and also receive the signal from the protons as the energy is released, providing the signal from which the image is created. The larger the RF coil, the less sensitive it is to the signal. However, generally speaking, the larger the RF coil the more uniform the excitation field. Innovations in RF coil engineering have mainly entailed decreasing the size of the coils and adding more coils around the head or body. If multiple receiver coils are added, typically a separate single RF coil is used for excitation. Recently, multiple coils have been used for excitation as well. Adding more coils has several advantages. First, sensitivity is increased since there are many sensitive small coils rather than one less sensitive large coil. Second, certain pulse sequences actually use the coil position to help create the image itself—allowing the data to be collected more rapidly. This improvement results in either higher resolution, more signal (as a shorter echo time or "TE" can be used when a shorter readout window is possible), or more rapid collection of whole-volume data sets.

By the early 2000s, eight channels—or separate coils—were added and then, by about 2006, sixteen and thirty-two channels were added. Currently, up to 128 RF coils have been used. In fact, a novel acquisition scheme known as "inverse imaging" (INI) has been used to virtually eliminate the need for spatial encoding gradients, thus substantially increasing the temporal resolution as well as decreasing the acoustic noise associated with acquisition, since gradient switching—essential for standard imaging formation—is what contributes to the high level of acoustic noise associated with MRI.

A second innovation in RF coils is on the excitation side. Technology for multichannel excitation has been in existence for over twenty years, yet only recently has it been used in conjunction with RF shimming methods that allow a uniform RF excitation profile to be obtained by adjusting each individual coil's power. This area is still in its infancy, since with multiple RF coils there is a danger of establishing RF "hot spots" in the brain where coil excitation fields overlap. Mitigation of this danger requires modeling of the head structure and accounting for any possible variation in head structure across individuals. This problem has not yet been fully resolved.

Gradient Coils

To review, the primary magnetic field B_0 causes the protons to precess at a specific frequency and is necessary for the creation of a "magnetic moment" that is the source of the signal. The oscillating RF field, B_1, is resonant with the precession frequency (in the RF range) and "excites" the protons. After a short time following the pulse of RF power, the protons start to "relax" and give off signal that is detected by the receiver coils in the transverse plane. At this moment, the signal needs additional "encoding" to be turned into an image. The basic principle at work is that the precession of the signal is directly proportional to the magnetic field that it is experiencing. In order to differentiate the signal in space, spatial gradients in the magnetic field have to be created rapidly, using a gradient coil. If a coil creates a gradient in space, we now have the protons precessing at a frequency and phase that is specific to their location in space. Once the signal is finally collected during spatial encoding, an algorithm known as the Fourier transform is applied to the data to create the image.

For high-speed imaging, the gradients need to perform this spatial encoding, line by line, in about 30 ms, implying that to create a 64 × 64 or 128 × 128 matrix they have to switch at a rate above 2 kHz or 2000 times per second. The technique that involves this rapid switching of the gradients is called echo planar imaging (EPI) and is the primary sequence used in all of fMRI. With each RF excitation and subsequent "echo," an entire plane or slice of data is created. The switching of the gradient induces large torque as it opposes the primary magnetic field, causing subtle yet loud mechanical vibrations. This is the source of the loud noise in MRI and fMRI. Typical clinical sequences collect data line by line using multiple RF pulses per plane, however EPI collects all the lines in one plane at once. When having EPI performed, the subject hears a "beep" sound for each image. These beeps are about 2 kHz.

In the very early days of fMRI, EPI could only be carried out at a resolution of about 3 mm^3 in two ways: The first was with the use of a "resonant" gradient amplifier system (i.e., ANMR) that was retrofitted onto an otherwise standard scanner. The second was with the use of low inductance, home-built head-only gradient coils. The home-built gradient coils helped deliver extremely high gradient switching rates (up to 200 mT/m/s) as well as very high gradients that are useful for high diffusion weighting. Once the major vendors implemented more powerful gradient amplifiers and other engineering improvements, local gradient coils were no longer necessary for performing EPI.

In recent years, head-specific gradient coils once again have started to make a small comeback as improved technology (hollow current-carrying wires for more efficient circulation of cooling fluid), increased need (even more gradient strength for faster imaging and higher resolution), and niche markets (head only) have entered the scene. While local gradient coils have disadvantages that include a level of added confinement, awkwardness for patients/subjects, and a more nonlinear gradient profile, they do have several advantages. Their lower inductance allows faster switching and higher gradients for a given current. Also, as the gradient falls off rapidly outside the coil, the change in magnetic field per unit time (dB/dt)— which increases with the distance of the gradient from the pivot axis isocenter—is not as high as with a whole-body gradient coil that creates gradients much further away from the pivot axis and does not cover such critical regions as the heart. It is understood that the biologic limits of dB/dt (or the rate at which the magnetic field changes in a specific location) are well below the physical limits imposed by the gradient coil. In other words, the coils are physically capable of switching much faster, but are limited by the fact that if they switch at a higher rate muscle twitching will be induced. However, if we limit the size of the gradient coil, it is possible to have a higher dB/dt at the point where imaging is taking place without being limited by the higher dB/dt further away from the pivot axis. In spite of this, it seems that in a clinical market, local coils will not be widely used—unless the research market changes such that high speed is clinically called for.

Pulse Sequences

A pulse sequence is the list of commands that are given to an MRI scanner to create an image. The details of pulse sequences and image formation are beyond the scope of this book; however, it's useful to have a general understanding and appreciation of what pulse sequences do because a fundamental aspect of fMRI method development is the advancement of pulse sequences for hemodynamic specificity, spatial or temporal resolution, sensitivity, or better brain coverage.

A pulse sequence generally involves a sequence of commands to the scanner to provide an RF pulse (to excite the protons so that there is signal to create an image), to create transient magnetic field gradients (to spatially encode the image), and to acquire the signal during these gradients.

Specific pulse-sequence timings (adjusting when the RF pulses, gradients, and image acquisition take place) have a profound effect on the type of image contrast that is created. Depending on the pulse sequence that is played out, an MR image can highlight white matter, gray matter, CSF, water diffusion direction, and many more tissue properties. The pulse sequence also determines the image resolution and acquisition speed. It is fundamentally important to understand that the aspect of resolution and speed—as with EPI—is relatively independent to the part of the pulse sequence that influences the functional contrast.

It's also important to discuss the EPI pulse sequence just a bit more. Because EPI collects an entire plane of data in less than 50 ms, all physiologic noise that can vary over time is "frozen" in time. Methods that take longer have an additional instability as the raw data for each image is collected across different times in cardiac and respiratory cycles, leading to "ghosting" artifacts that vary over time. EPI also has "ghosts", but they are stable over time as each echo planar image is collected in a short enough time to "freeze" any cardiac or respiratory changes during image collection. Because of this advantage, the temporal stability improvement is substantial as well as critical to fMRI for detecting very subtle 1%–5% signal deviations associated with blood oxygenation changes.

Almost every fMRI pulse sequence uses EPI for acquisition. However, the EPI gradient readout is only part of the pulse sequence. The sequence timing and RF pulses can be adjusted to create different contrast sensitivities. In the context of fMRI, pulse sequences can be crafted to create functional contrast weightings that can highlight large vessel flow, capillary perfusion, blood oxygenation, and blood volume. These contrast sensitivities were described in chapter 4.

In the early 2000s the inception of methods for using multiple independent receiver coils to improve spatial encoding efficiency made an impact in fMRI. As described earlier, imaging encoding, typically performed by gradients, can also be carried out by using multiple independent receiver RF coil sensitivity profiles—known as SENSitivity Encoding (SENSE). These have allowed either higher resolution imaging at a given readout window length or higher speed (i.e., shorter readout window length) imaging at a given resolution. In the context of fMRI and diffusion imaging, images of at least twice the resolution typically obtained have been collected using a single-shot approach. Since single-shot imaging is essential for the high stability needed in fMRI and diffusion tensor imaging (DTI), a method for performing higher resolution in a single shot was embraced. Now a large fraction of high-resolution fMRI studies is performed using single-shot SENSE acquisition methods.

A remaining problem when collecting high-resolution images is that with such thin slices, more slices and therefore more time are required to cover the entire brain per repetition time (TR). Assuming that about 15 images can be collected in a second, if 100 very thin slices are required to cover the entire brain, then the minimum repetition time (the time between each volume) or TR required to achieve whole brain coverage would be just under 7 sec. A TR of this length would typically make high-resolution whole-brain imaging prohibitively long because a minimum number of time points typically need to be collected—especially with less SNR at high resolution—to achieve statistical significance in brain activation images.

The answer to this problem arose independently from the Massachusetts General Hospital group[2] and the University of Minnesota group in collaboration with Berkeley.[3] The general concept is to use simultaneous multiplexed excitation of several imaging planes, allowing slice collection speedup by a factor of up to 8. This technique generally is known as simultaneous multi-slice (SMS) imaging. Instead of 15 slices per second, it is now possible to collect up to 120 slices per second using this approach!

The highest single-shot EPI *functional* resolutions ($<1mm^3$ voxel sizes) obtained have been in cortical layers and columns. Ocular dominance and orientation column activity have been mapped in humans.[4] Layer-specific mapping has also been accomplished. The ability to resolve layer-specific activity is of high potential significance as it is generally known that specific layers provide an "output signal" to other parts of the brain and other layers provide an "input signal." Teasing out this output/input directionality is a new frontier in fMRI that may shed light on what is understood about the functional circuitry of the healthy human brain.

In the direction of speed at perhaps the expense of resolution, a method involving using the RF coils almost exclusively for spatial localization, known as INI, has been developed.[5] Several groups have introduced echo volume imaging (EVI) as well.[6] This method allows collection of an entire volume of data in a single echo—as the name suggests. The major advantage is that motion correction works much better when the entire volume is a rigid body collected at one moment. In most cases with multi-slice imaging, through-plane motion is represented by a difficult-to-correct shearing across the volume as each slice is collected at a different time during the motion.

Pulse sequences can be designed and used to minimize time series noise. A recent example of this facet of innovation concerns multi-echo gradient echo EPI acquisition for fMRI time series. BOLD signal changes are fundamentally manifest as changes in $T2^*$ while most artifactual signal changes do not involve changes in $T2^*$. Change in $T2^*$ cannot be differentiated from non-$T2^*$ changes using a single echo for acquisition—as most fMRI time series obtain. Multi-echo acquisition, involving at least two readout windows per excitation along the T2* relaxation curve, allows baseline and activation-induced changes in T2* to be characterized and separated from artifact. Kundu et al. have developed an approach in which three-echo EPI time series are collected. [7] From this time series, independent component analysis (ICA) is applied. Each component can be analyzed across each TE to determine how well it fits into the T2* change curve. If the signal is BOLD based, the percent signal change will increase linearly with TE. If it is not BOLD, then the percent signal change will show no clear relationship to TE. The quality of fit for each ICA component is then calculated and ranked. From this ranking, a clear delineation of BOLD ICA components versus non-BOLD ICA components can be made.

A final pulse sequence innovation described here involves minimizing scanner acoustic noise. A major challenge in fMRI has been that the scanner is extremely loud, interfering not only with the presentation of acoustic stimuli but also with the interpretation of the acoustic-related brain activation. One answer to this problem was a development known as clustered volume acquisition. Rather than collecting an entire volume with each slice evenly spaced in time to continuously fill the acquisition time between volumes, the slices instead were "clustered" in time. For instance, if a TR were 4 sec, each volume would not take 4 sec to collect but rather, the entire volume would be collected in 1 sec, followed by 3 sec of silence. The TR itself was not altered, but now there was a period of silence that would allow subtle auditory stimulation.

Another potential solution to the challenge of acoustic noise is the use of silent pulse sequences. There are two strategies for producing less acoustic noise during MRI. The first strategy is to make more use of multiple RF coils to spatially encode the data than of the gradients. This approach, already mentioned, is known as INI (inverse imaging), and while it has limited spatial resolution, INI is virtually silent and extremely fast—allowing for sub-100 ms TRs (6). The second strategy is to use multi-shot sequences that involve slowly ramping the gradients and applying small RF pulses during the gradient application. The images produced are a bit noisier and less stable than EPI but that is because the gradients are not driven as hard as they are during typical imaging and therefore produce significantly less acoustic noise.[8]

MRI Acquisition-Related Issues

When starting an fMRI experiment, typical questions that must be addressed are: What resolution should I use? What TR should I use? How many images per time series should I collect? How many time series in the session should I obtain? How thin should my slices be? What do artifacts look like? The answers to these questions are all intertwined— a choice in one will influence the constraints of the others.

Each of the categories that follow are linked in many different ways. The following section informally walks readers through the trade-offs and issues involved with the practicalities that should be considered when performing an fMRI experiment.

Acquisition Rate

The single-image acquisition rate ultimately is limited by how fast the signal can be digitized and how rapidly the imaging gradients can be switched to create each line of raw data to then form into an image. MR imaging can be divided into single-shot and multi-shot techniques. Single-shot techniques typically are used for fMRI time-series collection as they are efficient and produce stable time series. Multi-shot techniques are for high-resolution structural scans used typically either for a structural reference for the lower resolution functional scans or for morphometric analysis of subject populations.

In single-shot EPI, the entire data set for a plane typically is acquired in about 20 to 40 ms. In the context of performing a BOLD experiment, the echo time (TE) or the center of the readout window will be about 20 to 40 ms as the optimal TE is equal to the T2* of the tissue. Along with some additional time for applying other necessary gradients, the total time for an image to be acquired is about 60 to 100 ms, allowing ten to sixteen images to be acquired in a second. Improvements in digital sampling rates and gradient slew rates will allow small improvements.

With multi-shot imaging, a single "line" or set of lines of raw data is acquired with each RF excitation pulse. Because of the relatively long time it takes for the longitudinal magnetization to return to equilibrium (characterized by the T1 of the tissue), a certain amount of time, between 50 and 500 ms, is spent waiting between shots. Otherwise, there would quickly be no signal left as the signal is not allowed to recover between excitation pulses—also referred to as becoming "saturated." Because of this necessary waiting time, multi-shot techniques generally are slower than single-shot techniques. For a 150-ms "waiting time" (or repetition time: TR), an image with 128 lines of raw data would take 150 ms × 128 = 19.2 sec, which is generally much too long for the collection of an fMRI time series.

As mentioned in the preceding section, improvements in the acquisition rate have been achieved with the use of parallel imaging techniques. Simultaneous acquisition of spatial harmonics (SMASH) was introduced in the early 2000s. The SMASH technique, introduced by Sodickson and co-workers,[9] uses linear combinations of coil signals from a surface coil array to replace time-consuming gradient steps. Following the introduction of SMASH, a second similar technique called SENSitivity Encoding was developed by Pruessmann and colleagues.[10] These approaches make use of the insight that multiple RF coils, having spatially separate sensitivity profiles, can aid in spatially encoding the data. SMASH does this encoding in raw data space, otherwise known as k-space, and SENSE does this in image space, working with data that has been reconstructed. Both methods can reduce acquisition time by a factor of 2 to 5 and are effective both in structural imaging and in fMRI acquisition.

Lastly, in the past five years, a new method has been introduced, allowing a large increase in the number of slices to be obtained per TR. Even with SENSE and SMASH for fMRI, a long echo time optimized for BOLD contrast (about 30 ms) must be used. Because of this, the number of slices per TR was still limited to about twenty or so. This approach uses the concept of simultaneous RF excitation of multiple slabs. A single composite RF pulse excites several slices at once, which are then unaliased (separated) during image reconstruction. These approaches are generally referred to as simultaneous multi-slice or multi-band techniques.[11]

Both SENSE and multi-band have been used—most notably in the Human Connectome data set.[12] The approach allowed for a time series of whole-brain EPI data to be obtained with a TR of less than 0.5 sec. Shortening the

TR is advantageous in time series acquisition as it enables more points in a time series for averaging as well as a finer temporal sampling—allowing more precise filtering out of noise. Therefore, it's almost always better to collect brain volumes as fast as possible.

Spatial Resolution

The spatial resolution in fMRI primarily is determined by the gradient strength, the digitizing rate, and the time available before the signal dies away. For multi-shot imaging, as high resolution as desired can be achieved if one is willing to wait while collecting lines of data with more RF pulses. For echo planar imaging, the signal decay rate (described by $T2^*$ with gradient echo EPI and by $T2$ with spin echo EPI) plays a significant role in determining the resolution. One can only sample for so long before the signal has completely decayed away. For this reason, echo planar images are generally lower resolution than multi-shot images. But with the development of the previously mentioned techniques, single-shot images can have voxel sizes as small as $1mm^3$, assuming there is enough signal to noise. Typically, fMRI time series have a temporal SNR of just over 100 to 1. This ratio is primarily determined by physiological noise. At lower SNR values, thermal noise is the dominant source of noise. If one wants to go to extremely high resolution, a price will be paid in SNR as the SNR is directly proportional to voxel volume. The lowest

SNR that one can practically use—with the limitation of time to average activations during a typical one hour session—is about 20 to 1.

Signal to Noise
The signal to noise and the functional contrast to noise are influenced by many variables. These include, among other things, voxel volume, echo time (TE), repetition time (TR), flip angle, receiver bandwidth, field strength, and RF coil used. Not considering fMRI for a moment, the image signal to noise is increased with larger voxel volume, shorter echo time, longer repetition time, narrow receiver bandwidth, higher field strength, and smaller RF coil. In the context of fMRI, the functional contrast to noise is optimized with a voxel volume equaling the size of the activated area, TE ≈ gray matter T2*, short TR (optimizing samples per unit time), narrow receiver bandwidth, high field strength, and smaller RF receiver coils. Smaller coils are now common in multi-channel arrays. These arrays typically have sixteen to thirty-two RF coils.

Stability
Theoretically, the noise, if purely thermal in nature, should propagate similarly over space and across time. In fMRI this is not at all the case since physiologic noise plays a large role over time-series data collection. For EPI, stability is much more of an issue on the longer time scale. Flow

and motion—both of which confound image quality—occur with cardiac and respiratory cycles. Subject movement and scanner instabilities also contribute. As mentioned already, single-shot acquisition such as EPI generally has better temporal stability than multi-shot techniques.

An SNR of 100 to 1 limit is determined mostly by physiologic noise as the brain is constantly pulsating with the heartbeat and breathing. Typically, it is optimal to adjust imaging parameters such that the image SNR matches the temporal SNR. If image SNR is higher than the temporal SNR, the time series is considered to be dominated by physiologic noise. Filtering out this noise has proven to be difficult, but in resting state fMRI, the noise can be used for the resting state fluctuation and connectivity information it contains. One of the most important and challenging problems in fMRI development work is the elimination of physiologic noise from the time series. If physiologic noise could be identified and effectively removed, then the temporal signal to noise would only be limited by the RF coil sensitivity, allowing fMRI time-series SNR ratios to approach 1000 to 1, opening up a wide range of applications and new findings using BOLD contrast. The concepts of image signal to noise versus time series signal to noise are illustrated in figure 13.

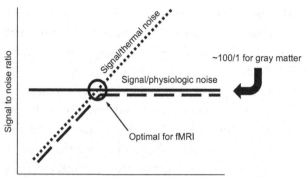

Figure 13 The signal to noise in fMRI time series. At low SNR values (left side of chart), thermal noise dominates, but at an SNR of approximately 100 to 1, physiologic noise, which also contains resting state fluctuations, starts to dominate. The dashed line is the time series SNR when collecting data from a living brain. Collecting EPI time series that have images with higher SNR than this physiologic noise limit does not add sensitivity. Therefore for fMRI, and in particular activation-based fMRI, it's optimal to collect data at the elbow of this dashed curve—where physiologic noise is just starting to impose its limiting effect. If there were no physiologic noise, then the temporal signal to noise would continue to increase as the image signal to noise increases (dotted line).

Image Quality

The most prevalent image quality issues are image warping and signal dropout. While books can be written on this subject, the description here is kept to the essentials.

Because much of image quality has to do with magnetic field inhomogeneity, it's useful to mention what is performed before each study to minimize this. Magnetic field "shimming" is a procedure in which current is adjusted through "shim" coils within the bore of the magnet that make small, spatially specific changes in the main magnetic field. This is a procedure in which specific areas where there is magnetic field inhomogeneity are targeted. The current in the shim coils is iteratively adjusted, typically using an algorithm rather than by hand, until the field inhomogeneities are reduced to a level that is satisfactory. That said, shimming is far from perfect, and field inhomogeneities still are prevalent, having a greater impact on image quality at higher field strengths and in particular on low-resolution, long readout-window sequences like EPI.

Image warping is fundamentally caused by three things: B_0 field inhomogeneities; gradient nonlinearities; and in the case where extremely large gradients are applied such as with diffusion imaging, eddy currents. A nonlinear gradient will cause nonlinearities in spatial encoding, causing the image to be distorted. This is primarily a problem when using local, small gradient coils that have a small region of linearity that drops off rapidly at the edges of the field of view. With the prevalence of whole-body gradient-coils for performing echo planar imaging, this problem is not a major issue. If the B_0 field is inhomogeneous, as is typically the situation with imperfect shimming procedures—particularly at higher field strengths, the protons will be processing at different frequencies than expected in their particular location. This will cause image deformation in these areas of poor shim—particularly with a long readout window or the long acquisition time of EPI. The long readout window duration allows for more time for these "off-resonance" effects to be manifest. Solutions to this include: obtaining a better B_0 shim, mapping the B_0 field to perform a correction based on this map, or, after the image has been reconstructed, performing image warping to match it with a non-warped high resolution structural image. Another viable solution is to reduce the readout window duration. This last solution is now possible with SMASH and SENSE imaging as this allows the same resolution image to be obtained in a fraction of the time, therefore producing images that suffer from much less warping.

Signal dropout is related to inhomogeneities in B_0, typically at interfaces of tissues having different susceptibilities. If within a voxel, because of the B_0 inhomogeneities, protons are precessing at different frequencies, their signals will cancel each other out. Several strategies exist for reducing this problem. One is, again, to shim as well as possible at the desired area. Due to imperfect shimming procedures, this solution helps but does not solve the signal dropout problem. The second potential solution is to reduce the voxel size (increase the resolution), thereby having less stratification of different frequencies within a voxel. The third potential solution is to choose the slice orientation such that the smallest voxel dimension (in many studies, the slice thickness is greater than the in-plane voxel dimension) is orientated perpendicular to the largest B_0 gradient or in the direction of the greatest inhomogeneity.

Lastly, gradient electronics and structural improvements have mitigated eddy currents for most imaging applications. However, when the gradients are pushed extremely hard in the case of diffusion imaging, eddy currents that transiently exist can occur during the readout window, distorting the images. To make the problem worse, the distortions will depend on the directions in which the diffusion gradients are applied. In the case of diffusion tensor imaging, where gradients are applied in many different directions, distortions will occur in many different directions, causing the images to be out of alignment in many areas. Spatial correction procedures exist to mitigate these issues, but they are not perfect.

This brings up a final important point. A common operation is to superimpose a functional image, obtained with an EPI time series, on top of a structural image, obtained with multi-shot acquisition. Because these sequences have different readout window widths, they will have different amounts of distortion—particularly in areas of poor shim that have large off-resonance effects. There are two solutions. The first is to perform in postprocessing a nonlinear image warping to better align the two images. This works for the most part; however, when attempting to align structure at the level of layers or columns, it tends to break down. The second solution is to use the EPI data for the structural underlay as well. For layer resolution work, this is essential as the EPI readout window is exceedingly long and the alignment has to be precise to less than about 0.1 mm. The general principle to take from this is that images with different readout window widths will have different levels of distortion that needs to be corrected if precise alignment is desired.

As with many of the topics discussed in this book, much more can be said, but the goal here is to introduce basic concepts and terms in a clear, practical way. MRI is a complex method that lies at the interface of engineering, physics, and human physiology. Development of all the technology mentioned in this chapter is still progressing—with improvements in speed, sensitivity, resolution, interpretability of the signal, and even the type of physiologic information obtained all coming about at a rapid rate.

MRI is a complex method that lies at the interface of engineering, physics, and human physiology. Development of all the technology mentioned in this chapter is still progressing.

STRATEGIES FOR ACTIVATING THE BRAIN

Fundamental to fMRI is the collection of a time series of whole-brain snapshots, spaced on the order of one second apart. The hemodynamic response to brain activation, while sluggish, is extremely well behaved and consistent. This consistency over time allows for a wide array of possibilities in designing a brain activation experiment.

The goal of designing activation paradigms is to localize and extract behaviorally relevant brain activation as efficiently and cleanly as possible. Paradigm design has been an area of abundant innovation. Most fMRI neuroscientists become adept at designing paradigms and tailoring them to the limitations and advantages of specific acquisition and processing strategies. This chapter highlights the types of paradigms that have been most novel and promising. These include blocked design,

Fundamental to fMRI is the collection of a time series of whole-brain snapshots, one second apart. The hemodynamic response to brain activation, while sluggish, is extremely well behaved and consistent. This consistency allows for a wide array of possibilities in designing a brain activation experiment.

event-related fMRI, phase-encoding, fMRI adaptation, resting state fMRI, naturalistic stimulus presentations, and real-time fMRI feedback. A background reference for this section is a paper by Amaro and Barker that summarizes study design and analysis in fMRI.[1] Figure 14 lists and shows iconic representations of seven classes of brain activation strategies that can be used.

Neuronal activation input strategies

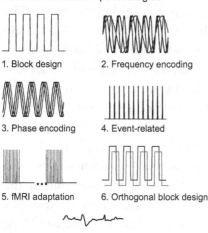

1. Block design

2. Frequency encoding

3. Phase encoding

4. Event-related

5. fMRI adaptation

6. Orthogonal block design

7. Free behavior design

Figure 14 An iconic depiction of key brain activation strategies used in fMRI. The most esoteric and least used is frequency encoding, which applies the concept of "multiplexing" multiple activation on-off frequencies. The event-related brain activation strategy, depicted here using constant intervals, now primarily employs jittered intervals to allow for deconvolution. Free behavior design appears to be gaining the most popularity.

Blocked Design

Functional MRI began with the use of blocked designs where subjects alternated approximately 10–30 sec of performing a task or receiving stimuli with an equal or slightly longer period of rest. A longer period of rest allows for the post-stimulus undershoot to return to baseline. A complete return to baseline can take up to 40 sec. Multiple tasks can be inserted as blocks. Images collected during each condition can be averaged and compared. The on-off switching in the signal allows separation of brain activation-related changes from the slow drifts that are ubiquitous in fMRI time series due to scanner instabilities or slow subject motion. Therefore, in order to optimally minimize the effects of low-frequency drifts, it is generally desirable to turn the tasks on and off as rapidly as possible and as many times as possible, pushing the on-off frequency as high as possible—separating it from low-frequency artifactual drifts—without compromising the amplitude of the response. The highest on-off frequency at which this can be obtained is about 10 sec on and 10 sec off or 0.05 Hz. The highest on-off frequency that is able to induce any measurable fMRI signal change is 0.67 sec on and 0.67 sec off or 0.75 Hz.

One useful advancement in blocked design paradigms was demonstrated by Courtney et al.,[2] in which six different time series related to working memory were efficiently incorporated into a single time series. The key aspect of this design was that the brain activation timing was such that the time series corresponding to each unique aspect of brain activity were all orthogonal to each other. By definition, all of the time series signals that are orthogonal to each other have a correlation of zero. These six regressors produced six different activation maps, all mathematically independent of each other. If paradigms are designed so that there is in fact some correlation between time series, time series can be mathematically "orthogonalized." However, this results in a loss of statistical power.

Interestingly, the 10 sec on and 10 sec off timing produces an activation time series that is nearly completely orthogonal to stimulus-induced motion. Such timing can be useful when having the subject perform tasks in the scanner that involve brief motion, such speaking words.[3]

It should be noted that much of the brain is behaving in what is likely a more temporally rich manner than what is typically modelled in a blocked design paradigm. Time-locked signal changes have been shown to vary considerably relative to a well-behaved on-off regressor. These signals are time locked, so they represent neuronal activity associated with the task, but are often overlooked by a choice of only a few model functions. Care should be taken when investigating the data to determine if any other repeatable, non-canonic signal changes have been missed.

While blocked designs are statistically the most powerful of paradigms, many brain activation tasks do not lend themselves to a constant duration of activity for tens of seconds. Event-related fMRI helps address these issues.

Event-Related fMRI

The idea of presenting and averaging multiple brief stimuli drew upon the vast EEG and MEG literature. In 1992, Blamire et al. first demonstrated event-related fMRI with a 3 sec visual stimulation.[4] Since they did not coin a term in that paper to describe this new paradigm, it is often overlooked. In 1996, Buckner et al. and McCarthy et al. first demonstrated event-related fMRI for cognitive tasks.[5] Other researchers have since demonstrated that there is practically no limit to how brief a brain activation duration is detectable with fMRI. The limitation was simply in how brief a brain response could be elicited. Stimuli as brief as 16 ms have elicited robust responses.

In the early days of event-related fMRI paradigm implementation, a brief task was given, and then there was a waiting period typically longer than 15 sec to allow the hemodynamic response to return to baseline.[6] This approach not only was statistically inefficient, it also was extremely boring for the subject and often induced undesirable "anticipation"-related activation as the subject awaited the next stimulus. During the mid- to late 1990s, an innovation—drawing on mathematics of linear systems—occurred in event-related fMRI. Researchers realized that the hemodynamic response could be modeled as a linear system. Neuronal inputs that are spaced such that their subsequent hemodynamic responses overlapped can be separated using simple deconvolution as long as the interstimulus intervals were varied or "jittered." Rapid event-related fMRI can typically accommodate an average stimulus interval of about 4 sec and stimulus duration of 0.5–3 sec, allowing a wide range and number of stimuli to be presented in a single time series. Studies followed that characterized the optimal average and distribution of interstimulus intervals, suggesting that, with event-related designs, the necessary timing for optimal creation of activation *maps* is very different than for characterizing the *shape* of the hemodynamic response from already known areas.[7]

Both blocked designs and event-related designs can be parametrically modulated, meaning that the task or stimulus intensity can be changed in a systematic manner such that it is reflected in the amplitude of the fMRI response. Varying an aspect of the task and then comparing the relative change in fMRI amplitude allows the precise determination of how much each region contributes to the corresponding processing. Some regions may not show a signal intensity modulation with task modulation but instead show a constant amplitude at all task intensities, while other regions may show a linear relationship. Others might show more complicated relationships. In general, parametric designs are common in fMRI as they powerfully differentiate the functional relationship between specific regions and task or stimulus quality or intensity.

Phase Encoding

Vision neuroscientists were among the first to embrace and advance the paradigm design methodology in fMRI. In 1994, Engel et al. were the first to demonstrate a continuous activation design that "phase-encoded" visual stimuli.[8] were the first to demonstrate a continuous activation design that "phase-encoded" visual stimuli. Here, the term "phase-encoding" means that the temporal "phase" corresponds to a specific visual-field location of the stimulus as well as a corresponding location in the brain in the case of retinotopy. The stimulus is never alternated with an "off" period, but rather, some aspect is varied continuously. In their study, a stimulus consisted of a flashing checkerboard ring that slowly increased in radius until it extended out to the periphery and then, once reaching the periphery, the ring repeated again, starting at

In general, parametric designs are common in fMRI as they powerfully differentiate the functional relationship between specific regions and task or stimulus quality or intensity.

the fovea. This cycle repeatedly activated a continuously varying ring of cortex. The second stimulus was a rotating wedge about 20 degrees in width and continuously rotated. The cycle time or the time for both approaches was about one minute, allowing for sufficient time for the hemodynamic response to keep up. A typical time series would be about five to ten minutes in duration, having five to ten phase-encoding cycles. Interestingly, this paradigm was able to delineate separate low-level visual areas as it is known that they are organized in a mirror-image fashion, such that the direction of activation associated with the moving wedge would cause opposite direction activation for each area; the borders of these areas can be determined if one pays attention to where the direction of activation changes.

Other applications of this type of paradigm have included: in the visual cortex, mapping spatial frequency selectivity; in the somatosensory cortex, mapping somatotopy; and in the auditory cortex, mapping tonotopy. In animal models, exquisite maps of cortical and subcortical auditory-frequency selectivity have been mapped using a paradigm in which the carrier frequency of a sound was varied slowly and continuously.

fMRI Adaptation

Grill-Spector and Malach originated "fMRI adaptation" paradigms where the amount of signal adaption with rapid sequentially presented stimuli is modulated by the degree to which neurons in each voxel consider the successive stimuli to be similar—either in spatial frequency or in semantic content.[9] Here, a stimulus or task is briefly given and then, at short intervals, either a similar or different stimulus or task is presented in a repetitive manner. If the same or similar stimulus is presented repetitively, then the underlying neuronal activity will quickly habituate and result in a blunted fMRI response. If different stimuli are presented, then a different pool of fresh neurons will respond and therefore the overall signal will not "adapt." This signal will have no subsequent adaptation with repeated stimuli. The assumption is that within each voxel there are different pools of neurons that respond differently to specific stimuli. These pools can be separated by their habituation to successive stimuli. This starts getting interesting when the stimuli are different yet share similar features. If neurons are sensitive to specific features of the stimuli only, then they will adapt when those features keep showing up even if the stimuli are all quite different.

Naturalistic Stimuli

A paradigm design trend that has been picking up momentum in the past decade presents naturalistic stimuli consisting of either movies, audio oration, or free self-paced multi-option behavior in the scanner. This approach then determines the precise correspondence of the time course of the fMRI signal change magnitude or connectivity with specific aspects of these stimuli or tasks as they are closely tracked and deconstructed.

Several analysis approaches have been used. The first is to create regressors corresponding to a specific aspect of the stimuli. In the case of watching a movie, separate regressors may show responses only when faces appear, or when people are talking. The regressors that are used could also be more specific, such as when subjects say a certain class of words, or gaze at the camera. After convolution with the hemodynamic transfer function, the calculations of the correlations between the regressors and the time series are carried out. Regions that show high correlation are inferred to be those activated by the specific aspect of the stimuli. The idea of creating regressors for as much from the stimuli as possible and then mapping the weightings of each of these onto the brain was coined "encoding" by Gallant who has implemented this approach in the context of visual, auditory, and semantic processing. The innovative processing approach that Gallant lab uses will be discussed more in the data processing section. The concept of "encoding" is related to "decoding" in that, for encoding, complex stimuli are broken down and the processing of the elemental parts are mapped onto the brain. For decoding, brain activation maps created by encoding are used as "training sets" that then can be used as templates to decode the pattern of brain activity with entirely novel stimuli—inferring what the subject was seeing, hearing, or doing based on the similarity of the pattern of activation.[10]

In the mid 2000s, the University of Pittsburgh, through the Organization for Human Brain Mapping (OHBM) and sponsored by Defense Advanced Research Projects Agency (DARPA), put on a competition in which entrants were challenged to deduce, based on brain-activation time courses, specifically what a subject was viewing, using training data sets in which the brain activation corresponding to a known movie stimulus was known. The winner was able to "decode" a movie, based on a previous training set of known movies, with extremely high accuracy from the brain activity alone.[11]

Hasson et al. advanced a different approach to analyzing naturalistic stimuli paradigms.[12] Their approach uses rich and varied yet identical stimuli and/or task timings across separate runs in a single subject or even across subjects such that the time course from one run can be used as the reference function for another run—assuming that the stimuli for all subjects are precisely time-locked with each other. This approach addressees a central problem prevalent in fMRI: "catching" all the interesting signals that arise in the data that may not resemble a canonical regressor shape. Any fluctuations in the signal that occur for the separate runs will be highly correlated, and therefore likely resulting from what was presented rather than from random fluctuations. Taken further, any differences in subjects' responses with identical time-series stimuli would imply a difference in how they process information.

Performing cross-subject correlation also helps eliminate spurious, non-task related temporal variations in the stimuli given that no two subjects are likely to have the same artifactual signal timings. Such a method has promise for effectively identifying salient differences and similarities across large numbers of subjects, and thus may be used to identify clinically relevant "biomarkers" corresponding to pathological conditions.

Resting State fMRI

Resting state fMRI is one of the most significant innovations since the inception of fMRI and is based on this observation: when the brain is not performing any overt task or receiving any stimulus, it is spontaneously active such that regions that are functionally related show correlation in their low-frequency (about 0.1 Hz) fluctuations—presumably driven by temporally correlated synchronous and spontaneous activity. The term "resting state fMRI" is a misnomer given the assumption that during "rest" the brain is not at all quiescent but rather shows constant churning activity that results in constant generation of fMRI signals. The temporal correlation of these signals from separate regions in the brain implies that they are "connected." Another name for resting state fMRI is functional connectivity fMRI. This name is problematic because many other variables than spontaneous activation can influence the degree of correlation in these time series signals.

A specific set of tools helped lay the groundwork for the discovery of the resting state fMRI innovation. In the software program AFNI, as well as its predecessors, users are able to choose a time course signal from the data—either from a voxel or from the average of multiple voxels—and then immediately use it as a reference function for comparison with all the individual voxel time courses in the brain. This tool set was extremely useful for exploring activation data in a rapid preliminary manner.

It was only a matter of time before someone would apply this tool to resting data where there was no obvious task. It took the curiosity of Bharat Biswal to use this tool in this manner.[13] His initial observation that the left and right motor cortex showed a high level of correlation even when the subject was not performing any overt task did not catch on immediately. His manuscript was published in 1995, yet the number of studies performing resting state fMRI were a mere trickle—accumulating no more than about five a year—until about 2005, when the field decided that resting state fMRI was real and useful for more than just mapping connectivity in the primary motor, visual, and auditory cortex. The rapid advance after this incubation period may also have been due to the fact that acquisition and processing methods had reached a state of sophistication such that detection of these resting state correlations was easy enough for a larger fraction of users to undertake. After about a decade researchers started to consider resting state studies that were complementary to their own activation studies, and then the floodgates opened as processing methods became more sophisticated. More interesting and encouraging results came out as more resting state networks were discovered. The number of networks or correlated nodes is not known, having increased in number—due to better processing methods and more sensitivity—from 5 or so upward to 350.[14] It's likely that the number of well-characterized nodes will continue to grow—perhaps until columnar organization is reached—as our ability to delineate fine detail increases.

Methods for de-noising, clustering, displaying, and comparing resting state data advanced after 2006. The field of resting state fMRI has helped spawn new journals such as *Brain Connectivity* and has become perhaps the dominant topic of study at the annual Organization for Human Brain Mapping Meeting. A large amount of federal funding has been redirected into understanding the human "Connectome" with resting state fMRI a major paradigm in this research.

Early on, and to some degree, still, it was thought that this signal was either an artifact of imaging, vigilance, or "vasomotion"—implying that the signal, while related to an underlying physiologic process, was not directly related to functional connectivity and therefore could not be reliably used as a measure of internal brain connectivity. The technique met skepticism as the fundamental observation of time series correlation between disparate and spatially independent signals in the brain had not been made in the context of other imaging modalities. It was almost too good to be true; however, as we are finding, it *is* true and works well.

Functional connectivity appears to be constantly changing over time during time series scans, suggesting a dynamic reconfiguration of networks or "brain state" changes. The ability to observe dynamic brain state changes has been brought about by a method by which the correlations of every region with every other region over

The field of resting state fMRI has helped spawn new journals such as *Brain Connectivity* and has become perhaps the dominant topic of study at the annual Organization for Human Brain Mapping Meeting. Federal funding has been redirected into understanding the human "Connectome" with resting state fMRI a major paradigm in this research.

a window of perhaps a minute or so in duration is carried out for each time point that is then "slid" along in time. This is known as the sliding window approach. In the presence of networks changing over time, it is a challenge even to really know what a network is. Some nodes jump from being correlated with one network to being correlated with another network. Which network does it belong to? A subfield of fMRI that focuses on dynamic connectivity changes has emerged, as it has been suggested that these dynamic states carry information about populations or individuals.

A central problem in all fMRI is how to cleanly separate meaningful fMRI signal from noise and artifact. This problem is much more difficult when studying resting state fMRI as there is no ground truth as to when the brain is active or not. Simultaneous external measures of neuronal activity, breathing, and cardiac function have can be used to help sort the fMRI signal but they are far from optimal. Currently, about half of the overall signal power that remains in the fMRI time series is not related to neuronal activity.[15]

Many open questions exist regarding resting state fMRI. Among the most pressing are the following:

1. What is the "purpose" of resting state fluctuations? What evolutionary need do they satisfy? What is the purpose of so much low frequency energy?

2. What temporal characteristics define each resting state network? Can they be defined are characterized in this way at all?

3. What about latencies? Accounting for hemodynamic latency variations across the brain may increase detectability and interpretability of resting state fMRI data. Do latencies contain more biologic meaning than simply downstream "draining vein" effects?

4. How many brain "states" exist? What are the time constants or sequence of these states? Do these measured states correspond to actual states of mind? Are they sensitive to pathology?

5. How finely delineated are these networks?

6. How sensitive is the resting state signal to pathology and prediction of treatment?

7. How much of the resting state signal is hemodynamic (i.e., vasomotion) versus neuronal in origin?

8. To what predominant neuronal oscillation frequency does the resting state fMRI signal correspond?

9. How is connectivity truly represented through correlation? Also, changes in connectivity typically are inferred by changes in correlation. A change in correlation—in the presence of noise—could also be

brought on by a simple change in amplitude of one signal or the other, or a simple change in SNR of one signal or the other.

Tenth, and last, a controversy that has been debated over the last few years concerns whether to regress out the global signal or not in order to increase sensitivity.[16] The growing consensus is that we should not remove the global signal as it artificially induces negative correlations between previously uncorrelated regions—thus complicating interpretation. However, a recent study has shown that this global signal is highly sensitive to the state of vigilance and removing it may normalize for vigilance variations across subjects and over time.[17] The question remains, what should we do about the global signal? It appears to depend on what we want to do with the time series. If we want to reduce vigilance-related variability across subjects, we might want to regress it out—but with caution. If we want to characterize differences in vigilance or vigilance-related processing across subjects, we might want to keep it in.

The growth of resting state processing methods is still accelerating. There will not be a single best way to analyze resting state data, but rather an array of approaches that give the researcher a tool set to address a particular question.

Real Time fMRI

Real time fMRI is the ability not only to collect and display the echo planar images during time series collection but also to continuously perform updated functional MRI analysis on the time series and to display this functional data to the investigator, or in some cases, back to the subject—allowing for image-based biofeedback. Real time fMRI was first developed in the mid-1990s by Cox, Jesmanowicz, and Hyde.[18] At this time, it pushed the limits of the available computation speed. Since then, with the advancement of computation speed and memory, having the ability to reconstruct and perform data analysis in real time is more commonplace. At many centers, real time fMRI has served for fMRI demonstration purposes. Showing brain activation as it happens in individual subjects during a scan can convey a powerful message to nonexperts that fMRI is indeed real. It gives a profound sense of the method's capacity to image activation as it is occurring.

However, real time fMRI has a much more important role. First, for fMRI to be used clinically, it is absolutely necessary that a technologist or physician be able to assess the data's quality. Even more than conventional MRI, motion can completely corrupt a data set. Often, motion artifacts cannot be corrected in post-processing, so the only reasonable option is to rescan if a subject or patient moves significantly. No clinical procedure would be workable—especially something as expensive as fMRI—with even a 10% failure rate. Therefore, being able to see the data and then run a rescan during the same session is very important for clinical applications.

Second, real time fMRI has been used to communicate with or assess locked in or otherwise uncommunicative patients. Rainer Goebel's group in Maastricht have developed methods for encoding letters in specific activation types.[19] They have also developed, using real time fMRI with feedback to the subject, the ability for a subject to move a video game paddle with a specific set of thoughts. They have demonstrated that two individuals, each being scanned, can play "pong" with only their thoughts controlling the paddles. Owen has been able to demonstrate, using real time fMRI, conscious thought processes and memory recall of subjects that are locked in.[20]

Compelling recent work has demonstrated that fMRI has potential as a therapeutic tool. DeCharms has demonstrated innovative work involving reducing chronic pain in subjects.[21] Here, the main regions involved with pain perception first were mapped, and then the amplitude of the signal in these regions was immediately displayed to the subject. The simple instruction given to the subject was to reduce the signal amplitude by whatever mental strategy possible. After several sessions, subjects began to experience some success. These feedback-based mental strategies were correspondingly successful in reducing the pain. In addition, the ability for the subject to remember the learned strategy for reducing the pain intensity continued for several months after these procedures.

Clinical trials that involve the use of real time fMRI feedback to help alleviate depression have begun.[22] In these experiments the networks for emotion are mapped and the subject is then instructed to upregulate the fMRI signals associated with the emotional network. After several sessions, many subjects have figured out to upregulate this network, and consequently, feel less depressed.

So far, real time feedback fMRI is still in its infancy as we continue to learn the best feedback methods and timings, the optimal networks to modulate in order to modulate subsequent behavior, and the types of behavior that actually can be modulated. The potential therapeutic implications of resting state fMRI are substantial; however, much work needs to be done to introduce it into clinical practice.

fMRI PROCESSING

While understanding where fMRI contrast comes from and the pulse sequences and equipment necessary is fundamental, it can be argued that processing fMRI data is the most critical, diverse, interesting, open-ended, and challenging aspect of fMRI. Functional MRI involves several processing steps including taking the raw data coming off the scanner and transforming it into a time series of images, attempting to "clean up" the time series by performing motion correction as well as artifactual signal regression, and then extracting and displaying meaningful brain activation information from analysis of changes in the time series signal. The topic of raw image creation or "reconstruction" is beyond the scope of this book and is only mentioned briefly. However, pre-processing of the time series and processing methods for the generation of meaningful brain activation maps are covered in more

Functional MRI involves several processing steps including taking the raw data coming off the scanner and transforming it into a time series of images, attempting to "clean up" the time series by performing motion correction as well as artifactual signal regression, and then extracting and displaying meaningful brain activation information from analysis of changes in the time series signal.

detail. As a reference, figure 15 shows the basic processing steps involved with fMRI. The far-right column shows the possible directions in which to take the processing.

Image Reconstruction

Image reconstruction was mostly covered in chapter 5. It is typically performed during the scan with the vendor-provided software. Images typically come up on the screen as they are being created. After the raw images are created, they usually are saved in one of several standard formats, with Digital Imaging and Communication in Medicine (DICOM) being the most popular and standard. After this, the fMRI scientist downloads the time series data from the scanner and puts them in a format, such as the "nifty" format, which can be manipulated by one of the common processing packages such as AFNI, SPM, FSL, or Brain Voyager. The processing methods that follow work with the raw images coming off the scanner after they have been formatted for time series analysis.

Pre-processing of the fMRI Signal

After reconstruction, most but not necessarily all of the following steps are carried out: motion correction,

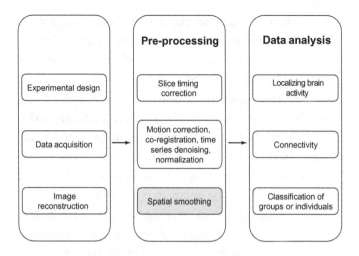

Figure 15 The fMRI analysis pipeline. Note that spatial smoothing is mostly relevant as a step for performing multi-subject averaging. It is not a recommended step when performing single subject analysis because a considerable amount of useful spatial functional information is lost in this process. Directions in which to take fMRI include simple localization and comparisons of localization differences. New directions include assessing connectivity changes, and assessing or predicting group or individual responses.

slice-timing correction, spatial filtering, temporal filtering, global intensity normalization, and spatial registration.

Functional MRI is highly sensitive to motion. Adjacent voxels can differ substantially in signal intensity. If one voxel moves into the other voxel's space even a small amount, the signal intensity change in both voxels will be large—ether positive or negative, depending on the motion direction and spatial signal intensity gradient. While motion can be manifest as a clear edge around steep signal intensity gradients (e.g., at the edges of the brain), it can also have profound effects throughout the brain. It is preferable to remove the effect rather than simply recognize it. Image registration involves aligning all images in the time series to the first image, a middle image, or the mean image. This helps to eliminate drift or many types of motion. Further motion correction can involve the modeling of motion based on the parameters extracted during registration as functions that can then be regressed from the data. This step is somewhat effective, but head movement introduces other artifacts that cannot be easily removed using such methods. If there is too much motion, these approaches will fail, requiring the scan to be performed again; however, this is often impossible or discovered after the patient or participant has completed the scan. Therefore, techniques for preventing or correcting head motion prospectively are important. Head movement can also be mistaken for task-related activation. Realignment can be confounded with functional activation and physiological noise. To help reduce these problems, several groups have implemented prospective motion-correction techniques based on optical estimates of motion and subsequent and very subtle readjustment of the imaging gradients.

In most fMRI, sequential volumes of data are collected using echo planar imaging. Each volume is made up of multiple slices or "planes." For most volumetric fMRI acquisition, each imaging plane in the volume is collected at a different time. A scanner may acquire planes in an interleaved, sequential, or center-out manner. Putting the slices back in order has been often done, but the benefits (in my opinion) of slice-timing correction are minimal, as this time difference is small relative to the hemodynamic response function as well as the spread of hemodynamic response lags across the brain.

Often spatial smoothing is performed with the idea that it increases image signal to noise by lowering resolution—given that the image signal to noise is directly proportional to voxel volume. Spatial smoothing is also often performed to match the spatial smoothness of an activation map with the variance that occurs with spatial normalization for multi-subject averaging. Typically, it is performed with a Gaussian kernel in 3D. Two disadvantages in smoothing are that the spatial and temporal structure is sometimes quite complex, and smoothing reduces this potentially interesting information. Small, significantly activated areas may be "smoothed out." It's also typically better from a signal to noise (higher SNR) and image distortion standpoint (less distortion with lower readout window) to collect at the resolution that one intends to use for analysis. Some groups are developing adaptive smoothing algorithms, in which the kernel varies as a function of the estimated intrinsic smoothness of the signal across the brain, but it may be advantageous to avoid smoothing altogether for most applications as there are many better ways to increase SNR with fewer disadvantages.

The fMRI time-series signal has significant non-neuronal fluctuations that span the entire frequency range. Thermal noise is present across all frequencies. Slow or low frequency drift due to slow settling of the head into the cushions or scanner instabilities is common. High frequency noise with turbulent blood and CSF flow and motion is also common. Temporal filtering is a common approach that potentially reduces these effects. High pass filters (i.e., those that only allow high frequency oscillations through) remove slow drift. Orthogonalization to a slow drift function also effectively removes slow drift—however, of course, if a task is performed or varied slowly, the signal corresponding to the slowly changing brain activation may also be removed by such slow drift removal methods. It is for this reason that brain activation is typically modulated relatively rapidly in an experiment (e.g., 10 sec on/10 sec off).

Global intensity normalization is aimed at correcting for run-to-run changes in signal intensity. This variance in signal intensity can often occur when auto pre-scan is allowed to repeat (and thus reset) at the beginning of each run, settling on slightly different values for excitation and receive gain for each run. To minimize cross-run variance, it's essential to keep one pre-scan setting. Even when one pre-scan is performed, cross-run variance can occur. Scaling every time series to a global mean eliminates this signal intensity shift across runs.

Other attempts at cleaning up the time series have also evolved over the years, including separately measuring the signal from obviously artifactual regions such as the sagittal sinus or ventricles, and regressing out this artifactual (and highly pulsatile) signal in various ways; collection of physiologic information such as heart rate, breathing, heart-rate variability, and breathing variability, and regressing this modeled signal from the time series; data "scrubbing" which involves removing time points identified as artifact; activation time-series timing adjustment to optimally separate the rapidly changing stimulus-correlated noise (perhaps from speaking in the scanner) from the slower activation-induced hemodynamic signal; and finally multi-echo acquisition to separate BOLD-like signal that shows a linear echo time (TE) dependence from non-BOLD-like signal that does not show a clear TE dependence.

Multi-echo EPI acquisition and analysis is a promising approach for effectively separating non-BOLD from BOLD signal in the time series. With one excitation, multiple EPI readout windows can be strung along during the signal decay. Groups have used these additional time series images at different TEs to increase statistical power and to increase sensitivity to areas that suffer from susceptibility signal dropout as shorter echo times have less susceptibility-related signal dropout. Multi-echo EPI also allows separation of activation-induced changes in $T2^*$ from changes in T1 or S_0. A simple voxel-wise fit of the $T2^*$ decay curve characterized by the multiple echoes allows for a quantitation in the change of the $T2^*$ decay rate as well as any change in the intercept of this fitted curve. The intercept represents the proton density, T1, or inflow effects—basically anything but transverse relaxation.

Basic Univariate and Multivariate Analysis

After the pre-processing steps, statistical analyses are carried out to determine which voxels are "activated" relative to a baseline or control condition. Significantly activated voxels are then typically color coded with a level of significance or the magnitude of activation and superimposed on the anatomic structural image, which is either a high-resolution MRI or one of the EPI scans from the time series. This analysis may include correlation analysis or advanced modeling of the expected brain activation. Statistical corrections can also be included, such as correction for smoothness of the measured time series at each voxel (if a time series is smooth, then there are less temporal "degrees of freedom") or corrections for nonindependence of adjacent voxels (if an image is smooth then each adjacent voxel has overlapping signal and therefore there are fewer spatial degrees of freedom). The main output from this step is a statistical map that indicates those points in the image where the brain has activated in response to the stimulus.

It is most common to analyze each time series in each region or voxel independently ("univariate analysis"). For example, standard GLM (general linear model) analysis is univariate. However, there are also "multivariate" methods that exploit the spatial relationships or relative *patterns* of activation. Most model-free methods are also multivariate. Multivariate approaches are powerful because they draw upon multiple "features" in the signal rather than just one. Multivariate methods boost sensitivity and offer new directions to take individual subject assessment. Figure 16 shows the basic concept of multivariate analysis and its ability to separate and classify individual subject data cleanly. It also show that any one variable is insufficient to cleanly separate these two populations; however, the simultaneous assessment of two variables or two "dimensions" allows clean separation. Multivariate analysis can involve a multitude of dimensions, but there is also a risk of "overfitting"—in which noise or idiosyncratic characteristics of the stimuli are used as a dimension, potentially implying a difference where none exists.

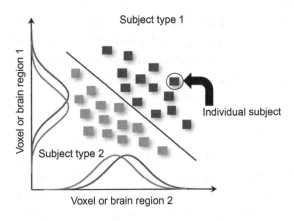

Figure 16 Two-dimensional multivariate assessment for individual subject classification. Each axis represents a measure. It might be the magnitude of signal from a voxel or region for each subject. Taken individually, no clear separation of the subject populations is obtained, but taken as a whole a "hyperplane" can cleanly separate the two populations, making individual assessment (answering the question if an individual belongs in one group or another) more tenable. In this manner, fMRI might be used to diagnose patients. Clinics may have n-dimensional "biomarkers" that are compared against patient data, offering valuable information related to how the patient compares with known populations.

It is important to differentiate between model-based and model-free methods. In a model-based method, a model of the expected response is generated and compared with the data. In a model-free method, aspects the data are found on the basis of their temporal or spatial or temporal independence. The researcher has to then decide whether or not the observed signal changes are neuronal or not. An example of model-free analysis is independent component analysis (ICA), which breaks the signal down into spatially independent maps that have unique time series. From these maps and time series, true activation may be discerned. However, the final decision of what constitutes "true" activation following this processing step is still an imperfect process.

General linear modeling sets up a model (i.e., a reference function) and fits it to the data. If the model is derived from the timing of the stimulation that was applied to the subject in the MRI scanner, then a good fit between the model and the data imply that the signal changes were probably caused by the stimulation or activation. The GLM normally is used in a univariate manner. A typical model function is created by starting with a stimulus function, which approximates the underlying hypothesized neuronal activity. This function is convolved with the hemodynamic response function (HRF)—representing a waveform approximate to what the BOLD impulse response should resemble. In the GLM, a matrix is set up in which each column represents a model or reference function.

To determine regions of activation, thresholding needs to be applied. The simplest method of thresholding is to select a statistically significant threshold and apply it to every voxel. However, because this operation is being performed on all voxels, one for each voxel in the brain, there are many tests being carried out. If 30,000 voxels are tested for a significance of $p < 0.01$, then it is expected that 300 will activate by chance, even if no stimulation is applied. A correction is necessary to reduce the number of false positives. Initially a Bonferroni correction had been used, where the significance level at each voxel is divided by the number of voxels. However, it was found that this threshold is far too conservative because activation is smooth in space and is composed of networks rather than independent voxels. Popular approaches have included Gaussian Random Field theory-based correction, statistical non-parametric mapping, and false discovery rate correction. Cluster extent-based correction has become popular due to its increased sensitivity. Current approaches focus on increasing power while retaining the ability to make specific inferences; however, further advances are likely possible only if prior information can increasingly be incorporated in quantitative, systematic ways to reduce the multiple comparisons problem.

It is common to run an experiment several times, either on the same subject or with several different subjects. Collection of multiple identical time series increases the sensitivity of the overall experiment or allows generalization of any conclusions to the whole population.

In order to combine statistics across different sessions or subjects, the first necessary step is to align the brain images from all sessions into some common space. This is typically performed using generic registration tools and

It is common to run an experiment several times, either on the same subject or with several different subjects. Collection of multiple identical time series increases the sensitivity of the overall experiment or allows generalization of any conclusions to the whole population.

can be carried out either on the raw image data or on the statistical maps.

Once all the data are aligned, a variety of methods can be used to combine results across sessions or subjects, to either create a single result for a group of subjects or compare different groups of subjects. These methods include "fixed-effects" and "mixed-effects" analyses. A fixed-effects analysis assumes that all subjects activate equally. It is only concerned with within-session and within-subject noise. A mixed-effects analysis also considers between-subject and sometimes between-session noise and errors, and therefore makes fewer assumptions about the data. Mixed-effects models that treat a subject as a random effect can be used to make inferences about the population from which the group of subjects is drawn. Fixed-effects results cannot be generalized beyond the subjects studied, and nearly universally require a lower standard of evidence.

How many subjects are required, in order for multi-subject statistics to be sensitive and robust? The answer depends on the level of response to the stimulation, between-subject variability, and scanner characteristics. Good estimates of these are rarely available for new research studies. A smaller number of subjects is generally required for a fixed-effects group analysis than for a mixed-effects analysis, but the ability to generalize a fixed-effects finding to a population is lower than for a mixed-effects finding.

Pattern Information Analysis and Machine Learning

Conventional statistical analysis of fMRI data focuses on finding macroscopic brain regions that are involved in specific mental activities. To find and characterize brain regions that become activated as a whole, data are usually spatially smoothed and activity is averaged across voxels within a region of interest (ROI). These steps increase sensitivity to relatively large-sized activations but result in decreased sensitivity to fine-grained spatial-pattern information. In the past several years, there has been a growing interest in going beyond activation assessment and analyzing fMRI data for the information carried by fine-grained patterns of activity within each functional region. Pattern-information-based techniques can move the level of analysis closer to assessing a region's representational content instead of its general involvement in task processing. Virtually all brain regions—even those as small as a single voxel—contain many millions of neurons that may code for diverse types of stimuli and mental events. Therefore, a voxel may be activated by many different tasks, but as part of distinct neural circuits. Within a blob of activation, the pattern of activation may show a unique relationship to subtleties of the specific task—resulting in subtle peaks and valleys within the blob. For example, a large blob of activation may appear when a subject is viewing faces, but the pattern of activation within that blob may change in subtle ways with each unique face being viewed. To pull out that information, multivariate pattern information analysis is applied.

The origins of pattern effect mapping in fMRI can be traced back to a study by Haxby et al.[1] Here the authors did something unique: They kept looking at what appeared to be a null result and, in the process, started an entirely new subfield of fMRI. They had been studying object processing, mapping the regions activated when volunteers viewed faces or houses. It appears that processing faces is so important to humans that it takes up a large swath of cortex, resulting in a clear "blob" of activation. Most reporting of fMRI results up to this point involved conveying the center of mass coordinates of blobs of activation. The field of fMRI has been fortunate that most activated regions were somewhat blobby and large, lending themselves to easy averaging across subjects and visualization with glass brains. The macroscopic systems-level organization of the brain was being fully studied and mapped, until Haxby et al. tried looking at objects with less evolutionary importance such as scissors versus shoes. On presenting these stimuli and analyzing the results, no clear blob—even a small one—distinguished itself. Instead, what the researchers saw was a "salt and pepper" pattern of activation that was, to the eye, not really different between the two less "important" or less clearly categorized stimuli. So, instead of moving on to something else, they calculated the activation maps for half of each data set and *calculated the spatial correlation between the two scissors maps and the two shoe maps and then between the scissors map and the shoe map* and so on. Their result was that even though the patterns looked like noise to the eye accustomed to seeing only clear blobs, the voxel-wise *patterns* of activation within the blobs clearly revealed themselves to be unique to the stimuli—showing much higher spatial correlation with the other half of the same data set than with another data set having a different stimuli. This study was likely the first to apply a rudimentary machine learning approach on a voxel-wise basis to pull out the unique voxel-wise pattern of activation associated with these subtler stimuli.

Many others have since picked up on this. Kriegeskorte, Goebel, and Bandettini developed a searchlight-based pattern-effect mapping.[2] Kamitani and Tong applied this to discerning which angle of grating a subject was viewing—even without their conscious knowledge.[3]

The discovery that fMRI can detect and map brain activity that is both subvoxel and uniquely distributed across many voxels is a leap of insight regarding not only what fMRI can detect but also how the brain is organized. New paradigms comparing the similarity of stimuli or tasks with the similarity of patterns of activity revealed a close connection between the two. Such analysis did not limit itself to voxels. Pattern similarity mapping has been carried out in the visual cortex of nonhuman primates (using implanted electrodes—each electrode serving as a "voxel") and compared directly with voxel-wise fMRI patterns in humans.[4] Such an avenue of investigation is literally wide open for new paradigms and new insights into similarities and differences across individuals rather than groups' averages. In fact, this approach defies spatial averaging of group maps as this fine-grained pattern effect would be washed away. A useful analogy would be if multiple photos of hands were averaged, the hand (activation blob) would remain while the fingerprints (pattern effect) would be smoothed away.

Further advances in brain reading, decoding, and encoding have been carried out, strongly suggesting that information—including object, word, sound, semantic content, and more—is distributed much more widely and finely than most researchers realized.[5] At this point in time, the field of fMRI decoding and encoding and multivariate assessment is relatively uncharted as these approaches allow us to look beyond blobs to subtler patterns that may reflect activation of subvoxel-scale neuronal populations.

The discovery that fMRI can detect and map brain activity that is both subvoxel and uniquely distributed across many voxels is a leap of insight regarding not only what fMRI can detect but also how the brain is organized.

Multimodal Integration

The integration of multiple imaging modalities has required innovation not only in how to physically collect the multiple different types of brain activation data simultaneously but also in how to process and interpret the results such that the whole is greater than the sum of the parts. One goal of multimodal acquisition has been to better understand the neural correlates of the fMRI signal. An approach to achieving this is to obtain simultaneous measures of underlying neuronal activity. So far, the techniques that have been used simultaneously with fMRI in humans have been EEG, near-infrared optical imaging, spectroscopy, electrocorticography (ECoG), and positron emission tomography (PET). Implanted electrodes,[6] calcium imaging methods, and optogenetic stimulation methods have been used in animal models. Scanners are being sold and used clinically that simultaneously collect PET and MRI data. Eye position, skin conductance, task performance, breathing rate, heart rate, and subtle external motion measures are collected simultaneously with fMRI to aid in processing and noise mitigation. Aside from the goal of better understanding the fMRI signal, EEG has been useful specifically for identifying stages of sleep as well as sub-ictal seizure activity, helping to identify time periods when the resting state signal shows meaningful differences in correlation. EEG has also proven extremely useful for assessing vigilance in resting state fMRI studies—which is a mostly overlooked but substantial confound for many studies (especially resting state) as subjects often fall asleep during scans, having a problematic effect on interpretation of final results.

Single Subject Classification

When fMRI first came about, results could be obtained on single subjects. A subject tapped their fingers and their motor cortex lit up—every single run, and for every single time that the subject tapped their fingers. It was easy to see. From the start fMRI has demonstrated clear results in individual subjects. As tasks became subtler and as the desire grew to compare populations, it became necessary to average multiple subjects together to see a significant effect. Therefore, for instance, differentiating brain activation as a function of reaction time might require an accumulation of multiple reaction times across hours of scanning—requiring multiple subjects not only to average together but to minimize individual idiosyncrasies. Another example: to understand the difference in brain activation to an n-back working memory task between subjects with a psychiatric disorder and subjects without would require multiple subjects in each group to determine the most prominent group difference.

While fMRI has been used clinically on an individual-subject basis for presurgical mapping, there is growing interest in applying fMRI prior to surgery to assess more than just the most strongly activated motor and sensory areas in the brain—so that a surgeon does not accidentally remove them. There's also a growing desire to use fMRI to help diagnose or monitor treatment for psychiatric or neurologic disorders as well as to help predict or measure responsiveness to specific therapies. So, while fMRI can readily differentiate group differences, its ability to determine with a high level of certainty which group an individual subject belongs to (normal, pathological, good/bad treatment outcome) is a much more difficult challenge to fulfill, as the two distributions of brain activation likely overlap considerably—each distribution having a wide spread corresponding to individual variability.

A promising strategy to deal with the challenge of individual subject assessment is the performance of multivariate classification in which multiple aspects of the data are assessed independently, as shown in figure 16. Rather than the magnitude of activation being compared in one region, perhaps the magnitude, latency, post-undershoot, and temporal standard deviation in multiple different regions for various segments of the task would be compared in a multidimensional space. If a clean difference (above 90% accuracy) is found that divides these variables in this multidimensional space, then the classifier is successful.

Each dimension might be magnitude of activation from a specific region or any other salient variable that has been determined to be a biomarker. If each distribution is considered alone, no statistical difference is obtained as there is too much overlap in each Gaussian distribution, but taken as a whole, a clear separation of all with a specific two-dimensional characteristic pattern is obtained. One can imagine considering a twenty variable or 20D classifier to perhaps separate normal subjects from those with a specific disorder. In this manner biomarkers can be developed in the future with fMRI to diagnose or at least complement other clinical assessments. In fact, data collected over the past twenty years could be reanalyzed for optimization of classification strategies. Regardless of what is carried out now, this approach requires very large data sets before a clear and robust classifier generally is successful and able to be distributed widely for diagnosis.

TWENTY-SIX CONTROVERSIES AND CHALLENGES

Functional MRI is unique in that, despite being an almost thirty-year-old method, it continues to progress in terms of sophistication of acquisition, hardware, processing, and our understanding of the signal itself. There has been no plateau in any of these areas. In fact, by looking at the literature, one gets the impression that this advancement is accelerating. Every new advance opens the potential range where fMRI might have an impact, allowing new questions about the brain to be addressed.

Despite its success—perhaps as a result of its success—fMRI has had its share of controversies coincident with methods advancements, new observations, and novel applications. Some controversies have been more contentious than others. Over the years, I've been following these controversies and have at times performed research to resolve them or at least better understand them.

Functional MRI is unique in that, despite being an almost thirty-year-old method, it continues to progress in terms of sophistication of acquisition, hardware, processing, and our understanding of the signal itself. There has been no plateau in any of these areas.

A good controversy can help to move the field forward, as it can focus and motivate groups of people to work on the issue itself, shedding a broader light on the field as these are overcome.

While a few of the controversies or issues of contention have been fully resolved, most remain to some degree unresolved. Understanding fMRI through its controversies allows a deeper appreciation for how the field advances as a whole—and how science really works—including the false starts, the corrections, and the various claims made by those with potentially useful pulse sequences, processing methods, or applications. Here is the list of twenty-six major controversies in fMRI—in approximately chronological order.

#1: The Neurovascular Coupling Debate

For nearly a century, the general consensus, hypothesized by Roy and Sherrington in 1890,[1] was that activation-induced cerebral flow changes were driven by local changes in metabolic demand. In 1986, a publication by Fox and Raichle challenged that view, demonstrating that with activation, localized blood flow seemingly increased beyond oxidative metabolic demand, suggesting an "uncoupling" of the hemodynamic response from metabolic demand during activation.[2] Many, including Louis Sokolof,

Understanding fMRI through its controversies allows a deeper appreciation for how the field advances as a whole—and how science really works—including the false starts, the corrections, and the various claims made by those with potentially useful pulse sequences, processing methods, or applications.

a well-established neurophysiologist at the National Institutes of Health, strongly debated the results. Fox nicely describes this period in history from his perspective in "The Coupling Controversy."[3]

I remember well, in the early days of fMRI, Dr. Sokolof standing up from the audience to debate Peter Fox on several circumstances, arguing that the flow response should match the metabolic need and there should be no change in oxygenation. He argued that what we are seeing in fMRI is something other than an oxygenation change.

In the pre-fMRI days, I recall not knowing in what direction the signal should go—such as when I first watched the impactful video presented by Tom Brady during his plenary lecture on the future of MRI at the Society for Magnetic Resonance (SMR) Meeting in August 1991; it was not clear from these time series movies of subtracted images the direction in which he performed the subtraction operation. Was it task minus rest or rest minus task? Did the signal go up or down with activation? I also remember very well, analyzing my first fMRI experiments, how I was expecting to see a decrease in BOLD signal—given that Ogawa, in an earlier paper,[4] hypothesized that metabolic rate increases would lead to a decrease in blood oxygenation and thus a darkening of the BOLD signal during brain activation. Instead, however, all I saw were signal increases. It was Fox's work that helped me to understand why the BOLD signal should *increase* with activation: flow goes up and oxygen delivery exceeds metabolic need, leading to an increase in blood oxygenation.

While models of neurovascular coupling have improved, we still do not understand the precise need for flow increases. First we had the "watering the whole garden to feed one thirsty flower" hypothesis, which suggested that flow increases matched metabolic need for one area, but since vascular control was coarse, the abundant oxygenation was delivered to a wider area than was needed, causing the increase in oxygenation. We also had the "diffusion limited" model hypothesizing that in order to deliver enough oxygen to the furthest neurons from the oxygen-supplying vessels, an overabundance of oxygen was needed at the vessel itself since the decrease of oxygen as it diffused from the vessel to the furthest cell was described as an exponential. This theory has fallen a bit from favor as the increases in $CMRO_2$ or the degree to which the diffusion of oxygen to tissue from blood is limited tend to be higher than physiologic measures. The alternative to feedback hypothesis involves neurogenic "feed-forward" hypotheses—which still doesn't get at the "why" of the flow response.

Currently, this is where the field stands. We know that the flow response is robust and consistent. We know that in active areas, oxygenation in healthy brains always increases, however, we just don't understand specifically why it's necessary. Is it neurogenic, metabolic, or some other mechanism to satisfy some critical evolutionary need that extends beyond the basic one for more oxygen? We are still figuring that out. Nevertheless, it can be said that whatever the reason for this increase in flow, it is fundamentally important, as the BOLD response is stunningly consistent.

#2: The Draining Vein Effect

"What about the draining veins?" I think this question was first posited at an early fMRI conference by Kamil Ugurbil of the University of Minnesota. Then and for the next several years he alerted the community to the observation that draining veins are a problem—especially at low field—as they smear and distort the fMRI signal change such that it's hard to know specifically where the underlying neuronal activation is with a high degree of certainty. In "20 Years of fMRI: The Science and the Stories," a special issue of the journal *NeuroImage*, Ravi Menon writes a thorough narrative called "The Great Brain versus Vein Debate."[5] When the first fMRI papers were published, only one, by Ogawa et al., was at high field (4 Tesla) and relatively high resolution.[6] Ogawa's paper showed there was a differentiation between veins (very hot spots) and capillaries (more diffuse, weaker activation in gray matter). Ravi followed this up with another paper using multi-echo imaging, to show that blood in veins had an intrinsically shorter T2* decay than gray matter at 4T and appeared as dark dots in T2*-weighted structural images, yet appeared as bright functional hot spots in the 4T functional images.[7] Because of the low SNR and CNR at 1.5T, allowing only the strongest BOLD effects to be seen, and because models suggested that, at low field strengths, large vessels contributed the most to the signal, researchers in the field worried that all fMRI was looking at was veins—at least at 1.5T.

The problem of large vein effects is prevalent using standard gradient echo EPI—even at high fields. Simply put, the BOLD signal is directly proportional to the venous blood volume contribution in each voxel. If the venous blood volume is high—as with the case of a vein filling a voxel—then the potential for high BOLD changes is also high if there is a blood oxygenation change in the area. At high field, indeed, there is not much intravascular signal left in T2*-weighted gradient echo sequences. However, the extravascular effect of large vessels still exists. Spin echo sequences (sensitive to small compartments) still are sensitive to the susceptibility effects around intravascular red blood cells within large vessels—even at 7T where intravascular contribution is reduced. Even with some vein sensitivity, promising high-resolution orientation-column results have been produced at 7T using gradient echo and spin echo sequences.[8] The use of arterial spin labeling (ASL) has potential as a method insensitive to large veins, although the temporal efficiency, intrinsic sensitivity, and brain coverage limitations blunt its utility. Vascular space occupancy (VASO), a method sensitive to blood volume changes, has been shown to be exquisitely sensitive to blood volume changes in small vessels and capillaries. Preliminary results have shown clear layer-dependent activation using VASO where other approaches have provided less clear delineation.[9]

Methods have arisen to identify and mask large vein effects—including thresholding based on percent signal change (large veins tend to fill voxels and thus exhibit a larger fractional signal change), as well as temporal fluctuations (large veins are pulsatile and thus exhibit more magnitude and phase noise). While these seem to be workable solutions, they have not been picked up and used extensively. With regard to using the phase variations as a regressor to eliminate pulsatile blood and tissue, it is likely that the primary reason for this not being adopted is that standard scanners do not produce these images readily, and therefore users do not have easy access to this information.

The draining vein issue is least problematic at voxel sizes larger than 2 mm because at these resolutions, regions of activation are typically "blobs" on the order of 1 cm in size. Other than enhancing the magnitude of activation, vein effects do not distort these "blobs" and thus typically are of no concern. In fact, the presence of veins helps to amplify the signal in these cases. Spatial smoothing and multi-subject averaging—still commonly practiced—also ameliorate vein effects as they tend to be averaged out as each subject has a spatially variant macroscopic venous structure.

The draining vein problem is most significant where details of high-resolution fMRI maps need to be interpreted for understanding small and tortuous activation patterns in the context of layer and column-level mapping. So far no fully effective method works at this resolution, as the goal is not to mask the voxels containing veins since there may be useful capillary and therefore neuronal effects still within the voxel. We need to eliminate vein effects more effectively on the acquisition side.

#3: The Linearity of the BOLD Response

The BOLD response is complex and not fully understood. In the early days of fMRI, it was found that BOLD contrast was both linear and nonlinear. The hemodynamic response tends to overestimate activation at very brief (<3 sec) stimulus durations[10] or at very low stimulus intensities.[11] With stimuli that were of duration of 2 sec or less, the response was much larger than predicted by a linear system. The reasons for these nonlinearities are still not fully understood. However, for interpreting transient or weak activations relative to longer duration activation, a clear understanding of the nonlinearity of neurovascular coupling across all activation intensities and durations needs to be well established.

#4: The Pre-stimulus Undershoot

The fMRI pre-undershoot was first observed in the late 1990s. With activation, it was sometimes observed that the fMRI signal, in the first 0.5 sec of stimulation, first deflected slightly downward before it increased.[12] Only a few groups were able to replicate this finding in fMRI; however, it appears to be ubiquitous in optical imaging work on animal models. The hypothesized mechanism is that before the flow response has a chance to start, a more rapid increase in oxidative metabolic rate causes the blood to become transiently less oxygenated. This transient effect is then washed away by the large increase in flow that follows.

Animal studies have demonstrated that this pre-undershoot could be enhanced by decreasing blood pressure. Of the groups that have seen the effect in humans, a handful claim that it is more closely localized to the specific regions of "true" neuronal activation. These studies were all in the distant past (more than fifteen years ago) and since then, very few papers have come out revisiting the study of the elusive pre-undershoot. It certainly may be that it exists, but the precise characteristics of the stimuli and physiologic state of the subject may be critically important to produce it.

While simulations of the hemodynamic response can readily reproduce this effect,[13] the ability to robustly reproduce and modulate this effect experimentally in healthy humans has proven elusive. Until these experiments are possible, this effect remains incompletely understood and not fully characterized.

#5: The Post-stimulus Undershoot

In contrast to the pre-simulus undershoot, the post-stimulus undershoot is ubiquitous and has been studied extensively.[14] Similar to the pre-undershoot, its origins are still widely debated. The basic observation is that following brain activation, the BOLD signal decreases and then passes below baseline for up to 40 sec. The hypothesized reasons for this include (1) a perseveration of elevated blood volume causing the amount of deoxyhemoglobin to remain elevated even though oxygenation and flow are back to baseline levels, (2) a perseveration of an elevated oxidative metabolic rate causing the blood oxygenation to decrease below baseline levels as the flow and total blood volume have returned to baseline states, and (3) a post stimulus decrease in flow below baseline levels. A decrease in flow with steady-state blood volume and oxidative metabolic rate would cause a decrease in blood oxygenation. Papers have been published arguing for and showing evidence suggesting each of these three hypotheses, so the mechanism of the post-undershoot, as common as it is, stays unresolved. It has also been suggested that if it is perhaps due to a decrease in flow, then this decrease might indicate a refractory period where neuronal inhibition is taking place.[15] For now, the physiologic underpinnings of the post-stimulus undershoot remain a mystery.

#6: Long-Duration Stimulation Effects

This is a controversy that has long been resolved.[16] In the early days of fMRI, investigators were performing the basic tests to determine if the response was a reliable indicator of neuronal activity. One of the tests was to determine if, with long-duration steady-state neuronal activation, the BOLD response remains elevated. A study by Frahm et al. came out suggesting that the BOLD response habituated after five minutes.[17] A counter-study came out showing that with a flashing checkerboard on for twenty-five minutes, the BOLD response and flow response (measured simultaneously) remained elevated.[18] It was later concluded that the stimuli in the first study were leading to some degree of attentional and neuronal habituation, not a long duration change in the relationship between the level of BOLD and the level of neuronal activity. Therefore, it is now accepted that as long as neurons are firing, and as long as the brain is in a normal physiologic state, the fMRI signal will remain elevated for the entire duration of activation.

#7: Mental Chronometry with fMRI

A major topic of study over the entire history of fMRI has been how much temporal resolution can be extracted from the fMRI signal. The fMRI response is relatively slow, taking about 2 sec to start to increase and, with sustained activation, about 10 sec to reach a steady-state "on" condition. On cessation, it takes longer to return to baseline—about 10 to 12 sec, and has a long post-stimulus undershoot lasting up to 40 sec. In addition to this slow response, it has been shown that the spatial distribution in delay is up to 4 sec due to spatial variations in the brain vasculature.

Given this sluggishness and spatial variability of the hemodynamic response, it may initially seem that there is no hope for subsecond temporal resolution. However, the situation is more promising than one would expect. Basic simulations demonstrate that, assuming no spatial variability in the hemodynamic response, and given a typical BOLD response magnitude of 5% and a typical temporal standard deviation of about 1%, after 11 runs of 5 minutes each, a relative delay of 50 to 100 ms could be discerned from one area to the next. However, the spatial variation in the hemodynamic response is plus or minus 2 sec depending on where one is looking in the brain and depends mostly on what aspect of the underlying vasculature is captured with each voxel. Large veins tend to have longer delays.

Several approaches have attempted to bypass or to calibrate the hemodynamic response. One way to bypass the slow and variable hemodynamic response problem is to modulate the timing of the experiment, so the relative onset delays with task timing delays can be observed. This uses the understanding that the hemodynamic response in each voxel is extremely well behaved and repeatable. Using this approach, relative delays of 50 to 500 ms have been discerned.[19] While these measures are not absolute, they are useful in determinating which areas show delay modulations with specific tasks. This approach has been applied—to 500 ms accuracy—to uncover the underlying dynamics and the relative timings of specific regions involved with word rotation and word recognition.

Multivariate decoding approaches have been able to robustly resolve subsecond (and sub-TR) relative delays in the hemodynamic response.[20] By slowing down the paradigm itself, Formisano et al. have been able to resolve absolute timing of mental operations down to the order of a second.[21] The fastest brain activation on-off rate has been resolved: it has recently been published by Lewis et al.[22] and is in the range of 0.75 Hz. While this is not mental chronometry in the strict sense of the term, it does indicate an upper limit at which high-speed changes in neuronal activation may be extracted.

#8: Negative BOLD Signal Changes

Negative BOLD signal changes were mostly ignored for the first several years of fMRI because researchers did not know precisely how to interpret them. After several years, with a growing number of negative signal change observations, the issue arose in the field, spawning several hypotheses to explain them. One hypothesis invoked a "steal" effect, where active regions received an increase in blood flow at the expense of adjacent areas that would hypothetically experience a decrease in flow. If flow decreases from adjacent areas, these areas would exhibit a decrease in BOLD signal but not actually be "deactivated." Another hypothesis was that these areas were more active during rest, thus becoming "deactivated" during a task as neuronal activity was reallocated to other regions of the brain. A third hypothesis was that they represented regions that were actively inhibited by the task. While in normal, healthy subjects the evidence for the steal effects is scant, the other hypotheses are clear possibilities. In fact, the default mode network was first reported as a network that showed consistent deactivation during most cognitive tasks.[23] This network deactivation was also seen in the PET literature.[24] A convincing demonstration of neuronal suppression associated with negative BOLD changes was carried out by Shmuel et al.[25] demonstrating, using a ring visual stimulus, a simultaneous decreased neuronal spiking and decreased fMRI signal in a ring of cortex that surrounds a ring of activation. These observations seem to point to the idea that the entire brain is tonically active and able to be inhibited by activity in other areas through several mechanisms. This inhibition is manifest as a decrease in BOLD.

#9: Sources of Resting State Signal Fluctuations

Since the discovery that the resting state signal showed temporal correlations across functionally related regions in the brain, there has been an effort to determine their precise origin as well as their evolutionary purpose. The predominant frequency of these fluctuations is in the range of 0.1 Hz, which was eye opening to the neuroscience community since, previously, most had not considered that neuronally meaningful fluctuations in brain activity occurred on such a slow time scale. The most popular model for the source of resting state fluctuations is that spontaneously activated regions induce fluctuations in the signal. As measured with EEG, MEG, or ECoG, these spontaneous spikes or local field potentials occur across a wide range of frequencies. When this rapidly changing signal is convolved with a hemodynamic response, the resulting fluctuations approximate the power spectrum of a typical BOLD time series. Direct measures using implanted electrodes combined with BOLD imaging show a temporal correspondence of BOLD fluctuations with spiking activity.[26] Recent work with simultaneous calcium imaging—a more direct measure of neuronal activation—has also shown a close correspondence both spatially and temporally with BOLD fluctuations,[27] thus strongly suggesting that these spatially and temporally correlated fluctuations are, in fact, neuronal. Mention of these studies is only the tip of a very large iceberg of converging evidence that resting state fluctuations are related to ongoing, synchronized, and spontaneous neuronal activity.

While the basic neurovascular mechanisms behind resting state fluctuations may be understood to some degree, the mystery of the origins and purpose of these fluctuations remains. To complicate the question further, it appears that there are different types of correlated resting state fluctuations. Some are related to the task being

Since the discovery that the resting state signal showed temporal correlations across functionally related regions in the brain, there has been an effort to determine their precise origin as well as their evolutionary purpose.

performed. Some may be related to brain "state" or vigilance. Some are completely insensitive to task or vigilance state. It has been hypothesized that the spatially broad, global fluctuations may relate more closely to changes in vigilance or arousal. Some are perhaps specific to a subject, and relatively stable across task, brain state, or vigilance state—reflecting characteristics of an individual's brain that may change only very slowly over time or with disease. A recent study suggests that resting state networks, as compared in extremely large data sets of many subjects, reveal clear correlations to demographics, life style, and behavior.[28]

Regarding the source or purpose of the fluctuations, some models simply state it's an epiphenomenon of a network at a state of criticality—ready to go into action. The networks have to be spontaneously active to be able to transition easily into engagement. The areas that are typically most engaged together are resultantly fluctuating together during "rest." In a sense, resting state may be the brain continually priming itself in a readiness state for future engagement. Aside from this issue there is the question of whether or not there are central "regulators" of resting state fluctuations. Do the resting state fluctuations arise from individual nodes of circuits simply firing on their own or is there a central hub that sends out spontaneous signals to these networks to fire? There has also been growing evidence suggesting that this activity represents more than just a subconscious priming. The default mode network, for instance, has been shown to be central to cognitive activity as rumination and self-directed attention.

Work is ongoing in trying to determine if specific circuits have signature frequency profiles that might help to differentiate them. Work is also ongoing to determine what modulates resting state fluctuations. So far, it's clear that brain activation tasks, vigilance state, lifestyle, demographics, disease, and immediately previous task performance can have an effect. There are at present no definitive conclusions regarding the deep origin and purpose of resting state fluctuations.

#10: Dead Fish (False Positive) Activation

In 2009, at the Organization for Human Brain Mapping meeting, Craig Bennett presented a poster on his study showing BOLD activation in a dead salmon's brain.[29] The poster was mostly a joke (I think), but with the intent to illustrate the shaky statistical ground on which fMRI stood with regard to the "multiple comparisons" problem.

Bennett's poster presentation was picked up by the popular media and was widely seen. His analysis of the dead fish's brain clearly suggested that fMRI based on BOLD contrast runs into some problems if it shows activation where there should be none. In fact, it was a clear indication of false positives that can happen by chance even if the appropriate statistical tests are not used. The basic problem in creation of statistical maps is that the maps not appropriately normalized to multiple comparisons. It's known that purely by chance, if enough comparison is made (in this case it's one comparison for every voxel), then some voxels will appear to have significantly changed in signal intensity. Bonferroni corrections and false discovery-rate corrections are almost always used and are available in current statistical packages. The Bonferroni correction method is likely too conservative as each voxel is not fully independent. The false discovery rate method is perhaps closer to the appropriate test. When using these two methods, false activations are minimized; however, they can still occur for other reasons. The structure of the brain and skull has edges that can enhance any small motion or system instability, resulting in false positives. While Bennett's poster brought out a good point and was among the most cited fMRI works in the popular literature and blogs, it failed to convey a more nuanced and important message: no matter what statistical test is used, the reality is that the signal and the noise are not fully understood, and therefore all are actually approximations of truth, subject to errors. That said, a well-designed study with clear criteria and models of activation as well as appropriately conservative statistical tests will minimize this false positive effect. In fact, it is likely that we are missing much of what is really going on by using oversimplified models of what we should expect of the fMRI signal.

A recent paper by Gonzalez-Castillo et al. showed that with a more open model of what to expect from brain activation, and nine hours of averaging, nearly all gray matter becomes "active" in some manner.[30] Does this mean that there is no null hypothesis? Likely not, but the true nature of the signal is still not fully understood, and both false positives and false negatives permeate the literature.

#11: Voodoo Correlations and Double Dipping

In 2009, Vul et al. published a paper that caused a small commotion in the field by calling out a clear error in neuroimaging data analysis and listing the papers—some quite high profile—that used this erroneous procedure, resulting in elevated correlations. The basic problem that was identified was that studies were performing circular analysis rather than pure selective analysis.[31] A circular analysis is a form of selective analysis in which biases involved with the selection are not taken into account. Analysis is "selective" when a subset of data is first selected before performing secondary analysis on the selected data. This is otherwise known as "double dipping." Because data always contain noise, the selected subset will never be determined by true effects only. Even if the data have no true effects at all, the selected data will show tendencies that they were selected for.

So, a simple solution to this problem is to analyze the selected regions using independent data (not data that were used to select the regions). Therefore, effects and not noise will replicate. Thus, the results will reflect actual effects without bias caused by the influence of noise on the selection.

This was an example of an adroit group of statisticians helping to correct a problem as it was forming in fMRI. Since the publication of their paper, the number of papers published with this erroneous approach has sharply diminished.

#12: Global Signal Regression for Time Series Cleanup

The global signal is obtained simply by averaging the MRI signal intensity in every voxel of the brain for each time point. For resting state correlation analysis, global variations of the fMRI signal are often considered nuisance effects and are commonly removed[32] by regression of the global signal against the fMRI time series. However, the removal of global signal has been shown to artifactually cause anticorrelated resting state networks in functional connectivity analyses.[33] Before this was known, papers were published showing large anticorrelated networks in the brain and interpreted as large networks that were actively inhibited by the activation of another network. If global regression was not performed, these "anticorrelated" networks simply showed minimal correlation—positive or negative—with the spontaneously active network. Studies have shown that removing the global signal not only induces negative correlation but also distorts positive correlations—leading to errors in interpretation.[34] Since then the field has mostly moved away from global signal regression. However, some groups continue to use it because it does clean up the artifactual signal to some degree.

The global signal has been studied directly. Work has shown that it is a direct measure of vigilance as assessed by EEG.[35] The monitoring of the global signal may be an effective way to ensure that when the resting state data are collected, subjects are in a similar vigilance state—which can have a strong influence on brain connectivity.[36] In general, the neural correlates of the global signal change fluctuations are still not fully understood, but it appears that the field has reached a consensus that, as a pre-processing step, the global signal should not be removed. Simply removing the global signal will not only induce temporal artifacts that look like interesting effects, but will also remove potentially useful and neuronally relevant signal.

#13: Motion Artifacts

Functional MRI is extremely sensitive to motion, particularly in voxels that have a large difference in signal intensity relative to adjacent voxels. Typical areas that manifest motion effects are edges, sinuses, and ear canals where susceptibility dropout is common. In these areas, even a motion of a fraction of a voxel can induce a large fractional signal change, leading to incorrect results. Motion can be categorized into task correlated, slow, and pulsatile. Work has been performed over the past twenty-seven years to develop methods to avoid or eliminate in acquisition or post-processing, motion-induced signal changes. In spite of this effort, motion is still a major challenge today. The most difficult kind of motion to eliminate is task-correlated motion that occurs when a subject tenses up or moves during a task or strains to see a display. Other types of motion include slow settling of the head during a scan, rotation of the head, swallowing, pulsation of blood and CSF, and breathing-induced motion-like effects.

Typical correction for motion is carried out by the use of motion regressors that are obtained by most image registration software. Ad hoc methods that can be effective for dealing with motion include visual inspection of the functional images and manually choosing time series signals that clearly contain motion effects that can then be regressed out or orthogonalized. Other approaches include image registration and time series "scrubbing." Scrubbing involves automated detection of "outlier" images and eliminating them from analysis. Other ways of working around motion have included paradigm designs that involve brief tasks such that any motion from the task itself can be identified as a rapid change whereas a change in the hemodynamic response is slow, thus allowing the signals to be separable by their temporal signatures.

In recent years, an effort has been made to proactively reduce motion effects by tracking optical sensors positioned on the head, and then feeding the position information back to the imaging gradients such that the gradients themselves are slightly adjusted to maintain a constant head position through changing the location or orientation of the imaging volume. The primary company selling such capability is KinetiCor.

A more direct strategy for dealing with motion is the implementation of more effective methods to keep the head rigid and motionless. This approach has included bite bars and plastic moldable head casts. These have some effectiveness in some subjects but run the risk of being uncomfortable—resulting in abbreviated scanning times or, worse, more motion due to active repositioning during the scan due to the subject's discomfort.

Aside from the problem of head motion, motion of the abdomen, without any concomitant head motion, can have an effect on the MRI signal. With each breath, the lungs fill with air, altering the difference in susceptibility between the signal in the chest cavity and the outside air. This alteration has an effect on the main magnetic field that can extend all the way into the brain, leading to breathing-induced image distortions and signal dropout. This problem is increased at higher fields where the effects of susceptibility differences between tissues are enhanced. Possible solutions are direct measurement of the magnetic field in proximity to the head using a "field camera" such as the one sold by the Swiss company Skope, and then perhaps using these dynamically measured field perturbations as regressors in post-processing or by feeding this signal to the gradients and shims prior to data collection in an attempt to compensate.

In resting state fMRI, motion is even more difficult to identify and remove as slow motion or breathing artifacts may have similar temporal signatures. Also, if there are systematic differences between intrinsic motion in specific groups such as children or those with attention deficit disorder (ADD), then interpretation of group differences in resting state fMRI results is particularly problematic as the degree of motion can vary with the degree to which individuals suffer from these disorders.

In high-resolution studies, specifically looking at small structures at the base of the brain, motion is a major confound because the base of the brain physically moves with each cardiac cycle. Solutions to this have included cardiac gating and simple averaging. Gating is promising; however, signal changes associated with the inevitably varying TR, which would vary with the cardiac cycle length, need to be accounted for by post-processing approaches that are often unreliable.

A novel approach to motion elimination has been the use of multi-echo EPI that allows the user to differentiate BOLD effects from non-BOLD effects based on the signal fit to a T2* change model.

Another motion artifact arises from through-plane movement. If the motion is such that a slightly different head position causes a slightly different slice to be excited (with an RF pulse), then some protons will not experience that RF pulse and will have magnetization that is no longer in equilibrium (achieved by a constant TR). If this happens, even if the slice position is corrected, there will be some residual nonequilibrium signal remaining that will take a few TRs to get back into equilibrium. Methods have been put forward to correct for this, modeling the T1 effects of the tissue. However, these can also eliminate the BOLD signal changes, so this problem still remains.

Lastly, apparent motion can also be caused by scanner drift. As the scanner is running, gradients and gradient amplifiers and RF coils can heat up, causing a drift in the magnetic field as well as the resonance frequency of the RF coils, which results in a slow shifting of the image location and quality of image reconstruction. Most vendors have implemented software to account for this, but this is not an ideal solution. It would be better to have a scanner that does not have this instability to begin with.

In general, motion is still a problem in the field and one that every user still struggles with. It is becoming better managed, however, as we gain greater understanding of the sources of motion, the spatial and temporal signatures of motion artifacts, and the temporal and spatial properties of the BOLD signal itself.

There's a payoff waiting once complete elimination of motion and non-neuronal physiologic fluctuations in solved. The upper temporal signal to noise of an fMRI time series is no higher than the ratio of about 120 to 1 due to physiologic noise setting an upper limit. If this noise source were to be eliminated, then the temporal signal to noise ratio would only be limited by coil sensitivity and field strength, perhaps allowing fMRI time-series SNR values to approach 1000 to 1. This improvement would transform the field.

#14: Basis of the Decoding Signal

Processing approaches have advanced beyond univariate processing that involves comparison of all signal changes against a temporal model with the assumption that there is no spatial relationship between voxels or even blobs of activation. Instead, fine-grained voxel-wise patterns of activation in fMRI have been shown to carry previously unappreciated information regarding specific brain activity associated with a region. Using multivariate models that compare the voxel-wise pattern of activity associated with each stimulus, detailed information related to such discriminations as visual angle and face identity as well as object category have been differentiated.[37] A model to explain these voxel-specific signal change patterns hypothesizes that while each voxel is not small enough to capture the unique pool of neurons that are selectively activated by specific stimuli, the relative population of neurons that are active in a voxel causes a modulation in the signal that, considered with the array of other uniquely activated voxels, makes a pattern that, while having limited meaningful macroscopic topographic information, conveys information as to what activity the functional area is performing. Another proposed explanation for the decoding signal is that it simply reflects subtle macroscopic changes that are occurring across the brain.

One study has attempted to answer the question of whether the source of pattern effect mapping is multivoxel or subvoxel by testing the performance of pattern effect mapping as the activation maps were spatially smoothed.[38] The hypothesis was that if the information were subvoxel and varied from voxel to voxel, then the performance of the algorithm would diminish with spatial smoothing. If the information was distributed at a macroscopic level, smoothing would improve detection. This study showed that both voxel-wise and multivoxel information contributed in different areas to the multivariate decoding success. Therefore, while the decoding signal is robust, the origins—as with many other fMRI signals—are complicated and varying.

#15: Signal Change but No Neuronal Activity?

Several years ago, Sirotin and Das reported that they observed a hemodynamic response where no neuronal activity was present. They recorded simultaneously hemodynamic and electrical signals in an animal model during repeated and expected periodic stimuli. When the stimulus was removed when the animal was expecting it, then there was no neuronal firing but a hemodynamic response remained.[39] Following this controversial observation, there were some papers that disputed their claim, suggesting that very subtle electrical activity was still present. The hemodynamic response is known to consistently overestimate neuronal activity for very low level or very brief stimuli. This study appears to be an example of just such an effect, despite what the authors claimed.

While there was no clear conclusion to this controversy, the study was not replicated. In general, a claim such as this is extremely difficult to make as it is nearly impossible to show that something doesn't exist when one is simply unable to detect it.

#16: Curious Relationships to Other Measures

Over the years the hemodynamic response magnitude has been compared with other neuronal measures such as EEG, PET, and MEG. These measures show comparable effects, yet there have been puzzling discrepancies reported. One example is the following paper,[40] comparing the visual checkerboard flashing-rate dependence of BOLD and of MEG. At high *spatial* stimulus frequencies, the flashing visual checkerboard showed similar monotonic relationships between BOLD and MEG. At low spatial frequency, the difference between BOLD and MEG was profound—showing a much stronger BOLD signal than MEG signal. The reasons for this discrepancy are still not understood, yet studies like these are important for revealing differences where otherwise it is easy to assume that they are measuring very similar effects.

#17: Contrast Mechanisms: Spin Echo versus Gradient Echo

Spin echo sequences are known to be sensitive to susceptibility effects from small compartments and gradient echo sequences are known to be sensitive to susceptibility effects from compartments of all sizes.[41] From this observation it has been incorrectly inferred that spin echo sequences are sensitive to capillaries rather than large draining veins. The mistake in this inference is that the blood within draining veins is not considered. Within draining veins are small compartments: red blood cells. So, while spin echo sequences may be less sensitive to large-vessel extravascular effects, they are still sensitive to the intravascular signal in vessels where the blood signal is present.

There have been methods proposed for removing the intravascular signal, such as the application of diffusion gradients that null out any rapidly moving spins. However, spin echo sequences have lower BOLD contrast than gradient echo by at least a factor of 2 to 4, and with the addition of diffusion weighting, the contrast almost completely disappears.

Another misunderstanding is that spin echo EPI is truly a spin echo. EPI takes time to form an image, having a long "readout" window—at least 20 ms whereas with spin echo sequences there is only a moment when the echo forms. All other times, the signal is being refocused by gradient echoes—as with T2*-sensitive gradient echo imaging. Therefore, it is nearly impossible in EPI sequences to obtain a perfect spin echo contrast. Most of spin echo EPI contrast is actually T2* weighted. "Pure" spin echo contrast is where the readout window is isolated to just the echo—only obtainable with multi-shot sequences. However, even at high field strengths, spin echo sequences are considerably less sensitive to BOLD contrast than gradient echo sequences.

There is hope that "pure" spin echo sequences at very high field might be effective in eliminating large vessel effects because at high field blood T2 rapidly shortens, and therefore the intravascular signal contributes minimally to the functional contrast. Spin echo sequences have been used at 7T to visualize extremely fine functional activation structures such as ocular dominance and orientation columns.[42]

For these reasons, spin echo sequences have not caught on except for a small handful of fMRI studies at very high field. If high field is used and a short readout window is employed, then spin echo sequences may be one of the sequences of choice for spatial localization of activation, along with VASO, the blood volume-sensitive sequence.

#18: Contrast Mechanisms: SEEP Contrast

Over the years, after tens of thousands of observations of the fMRI response and the testing of various pulse sequence parameters, some investigators have claimed that the signal does not always behave in a manner that is BOLD-like. One interesting example appeared about fifteen years ago. Here Stroman et al.[43] were investigating the spinal cord and failed to find a clear BOLD signal. On using a T2-weighted sequence they claimed to see a signal change that did not show a TE dependence and therefore was not T2 based. Rather, they claimed the signal change was based on proton density changes—but not perfusion. It was also curiously most prevalent in the spinal cord.

The origin of this signal has not been completely untangled and "signal enhancement by extravascular protons" or SEEP contrast has disappeared from the current literature. Those investigating spinal cord activation have been quite successful using standard BOLD contrast.

#19: Contrast Mechanisms: Activation-Induced Diffusion Changes

Denis LeBihan is a pioneer in fMRI and diffusion imaging. In the early 1990s, he was an originator of diffusion tensor mapping.[44] Later, he advanced the concept of intra-voxel incoherent motion (IVIM).[45] The idea behind IVIM is that with extremely low levels of diffusion weighting, a pulse sequence may become selectively sensitized to randomly, slowly flowing blood rather than free water diffusion. The pseudo-random capillary network supports blood flow patterns that may resemble, imperfectly, rapid diffusion, and thus could be imaged as rapid diffusion using gradients sensitized to this high diffusion rate of random flow rather than pure diffusion. This concept was advanced in the late 1980s and excited the imaging community in that it suggested that, if MRI were sensitive to capillary perfusion, it could be sensitive to activation-induced changes in perfusion. This contrast mechanism, while theoretically sound, was unable to be clearly demonstrated in practice as relative blood volume is only 2% and sensitization of diffusion weighting to capillary perfusion also, unfortunately, sensitizes the sequences to CSF pulsation and motion in the brain.

Le Bihan emerged several years later with yet another potential functional contrast that is sensitive not to hemodynamics, but rather—hypothetically—to activation-induced cell swelling. He claimed that diffusion weighting revealed that on activation, measurable decreases in diffusion coefficient occur in the brain. The proposed mechanism is that active neurons swell with neuronal activation, thus increasing the intracellular water content. High levels of diffusion weighting are used to detect subtle activation-induced shifts in neuronal water content. A shift in water distribution from extracellular space that has a slightly higher diffusion coefficient to intracellular space that has a slightly lower diffusion coefficient would cause an increase in signal in highly diffusion-weighted sequences.

While LeBihan has published these findings,[46] the idea has not achieved wide acceptance. First, the effect itself is relatively small and noisy, if present at all. Second, other published papers have demonstrated that the mechanism behind this signal is vascular rather than neuronal.[47] And third, and perhaps most important, many groups, including my own, have tried this approach, coming up with null results. If the technique gives noisy and inconsistent results, it likely is not going to compete with BOLD, regardless of how selective it is to neuronal effects. Of course, it's always worthwhile to work on advancing such methodology!

#20: Contrast Mechanisms: Neuronal Current Imaging

In the late 1990s several resarchers explored the possibility that MRI is sensitive to electric currents produced by neuronal activity. A current carried by a wire is known to set up magnetic fields around the wire. These magnetic fields when superimposed on the primary magnetic field of the scanner can cause MR phase shifts or, if the wires are very small and randomly distributed, an NMR phase dispersion—or a signal attenuation. In the brain, dendrites and white matter tracts behave like wires that carry current. Basic models have calculated that the field in the vicinity of these fibers can be as high as 0.2 nT.[48] MEG in fact, does detect these subtle magnetic fields as they fall off near the skull. At the skull surface the magnetic fields produced are on the order of 100 fT, inferring that at the source, they are on the order of 0.1 nT.

Over the last fifteen years, work has continued around the world toward the goal of using MRI to detect neuronal currents. This work includes attempts to observe rapid activation-induced phase shifts and magnitude shifts in susceptibility-weighted sequences as the superimposed field distortion would cause either a phase shift or dephasing, depending on the geometry. Using these methods, no conclusive results have been reported in vivo. Other attempted methods include "Lorentz" imaging.[49] Here the hypothesis is that when current is passing through a nerve within a magnetic field, a net torque is produced, causing the nerve to move just a small amount—potentially detectable by well-timed diffusion weighting. Again, no clear results have emerged. More recent approaches are based on the hypothesis that the longitudinal relaxation properties of spins may be affected if in resonance with intrinsic resonances in the brain such as alpha (10 Hz) frequencies.[50] Spin-locking approaches that involve adiabatic pulses at these specific frequencies aim to observe neuronal activity-based changes in the resonance of the NMR tissue. In such manner, maps of predominant oscillating frequency may be made. Again, such attempts have resulted in suggestive but not conclusive results.

Many researchers are working on methods to detect neuronal currents directly. However, by most calculations, the effect is likely an order of magnitude too small. Coupled with the fact that physiologic noise and even BOLD contrast would tend to overwhelm neuronal current effects, the challenge remains daunting. Such approaches would require extreme acquisition speed (to detect transient effects that cascade over 200 ms throughout the brain), insensitivity to BOLD effects and physiologic noise, yet likely an order of magnitude higher sensitivity overall.

#21: Contrast Mechanisms: MR Phase Imaging

The idea of observing phase changes rather than magnitude changes induced by oxygenation changes is not a new one. Back in the early 1990s researchers showed that large vessels reveal a clear MR phase change with activation. Several researchers suggested that this approach may allow the clean separation of large vessel effects from tissue effects when interpreting BOLD. In general, it is known that large changes in susceptibility relative to voxel size induce a net phase shift as well.

The concept of observing phase changes has been revisited. Studies have suggested that large activated regions while mostly showing magnitude changes may act as a single large susceptibility perturber and induce a large shift in phase. These studies suggest that use of both phase and magnitude information would boost sensitivity and specificity. This approach has not yet caught on in the field, possibly in part because vendors typically only provide magnitude reconstruction of the scanner data. Most users simply don't have access to NMR phase information. It may also turn out that any gains are very small, such that obtaining double the data and spending twice the time on pre- and post-processing are not worth the effort.

#22: fMRI for Lie Detection

Over the past twenty years, fMRI has provided evidence that the brain of someone who is lying is active in distinctly different ways than the brain of someone who is telling the truth. There is additional prefrontal and parietal activity with story fabrication.[51] There have been papers that have demonstrated this effect in many different ways. In fact, some studies show not only that lie detection is possible but also that extraction of the hidden knowledge of the truth is also possible using fMRI.[52]

Because of this success, companies touting MRI-based lie detection services have cropped up (e.g., No Lie MRI at http://noliemri.com/ and CEPHOS at http://www.cephoscorp.com/). The problem with this is that lie detection has never been fully tested in motivated and at times psychopathic criminals and negative results are more common than positive results. Inconclusive fMRI-based lie detection results, if allowed in court, could potentially bias in favor of the defense because a negative or inconclusive result would suggest that the individual is innocent.

The difference between research success and success or readiness for implementation in public use illustrates a problem that much of the field of fMRI faces. In real-world applications, there are many variables that make fMRI nearly uninterpretable. Generalizing from group studies or highly controlled studies on motivated volunteers to individual studies in patients or criminals is extremely challenging to say the least.

Nevertheless, in spite of scientific, legal, ethical, operational, and social hurdles, machine learning and classification methods may ultimately prove capable in interpreting individual subject activation in association with lie detection with actual criminals or other real-world situations. There may in fact be regions of the brain or patterns of activity that tell truth from lies regardless of whether the person is a psychopathic hardened criminal or a motivated college student volunteer. No one has done those comparison studies yet.

#23: Does Correlation Imply Connectivity?

In resting state and sometimes task-based fMRI, the correlation between voxels or regions is calculated. A growing trend is to replace the word "correlation" with "connectivity." The assumption is that a high temporal correlation between disparate regions of the brain directly implies a high level of functional connectivity. It's also assumed that any *change* in correlation between these regions implies that there is a corresponding change in connectivity. As a first approximation, these statements may be considered true, however there are many situations in which they are not true.

First, other processes can lead to a high correlation between disparate regions. Bulk movement, cardiac pulsation, and respiration are three primary sources of artifactual correlation. For the most part these are well dealt with—as motion is relatively easily identified and cardiac pulsation is at a distinct frequency (1 Hz) that fortunately is far from resting state correlation frequencies (0.1 Hz). However, aliased cardiac frequencies and respiration-induced correlations are a greater challenge to remove. Respiration also can create signal changes that show T2* changes, so multi-echo sequences optimized to separate BOLD from non-BOLD effects would be less effective in removing respiration effects. Respiration is identifiable by the fact that it is more spatially diffuse than correlations between distinct regions. This separation based on spatial pattern and spatial diffusivity is still extremely difficult to perform robustly and without error.

Modulations in correlation can occur for a number of reasons. First, when looking at a pair of regions that show correlation, if both signals contain noise (as they all do), an increase in the amplitude of one signal will naturally increase the correlation value, but no "real" increase in connectivity likely occurred. Likewise, if the noise in one or both of the signals is reduced, there will be a measurable reduction in correlation but probably no change in actual functional connectivity. If the relative latency or shape of one of the responses changes, then a change in correlation will occur, perhaps without a change in connectivity. Let's say an additional frequency was added from another oscillating signal that has nothing to do with the signal driving the "connectivity" between the two regions. If this happens, then again, the correlation between the two signals will be reduced without the connectivity between the regions being altered.

All of these issues have not been addressed in any systematic manner; we are still in the early days of figuring out the best ways to cleanly and efficiently extract correlation data. In the future, in order to make more meaningful interpretations of these signals, we will need to control for these potentially confounding effects.

#24: The Clustering Conundrum

In 2016, a bombshell paper by Eklund, Nichols, and Knuttson[53] identified an error in most of the statistical fMRI packages, including SPM, FSL, and AFNI. Quoting a part of the abstract of that paper: "In theory, we should find 5% false positives (for a significance threshold of 5%), but instead we found that the most common software packages for fMRI analysis (SPM, FSL, AFNI) can result in false-positive rates of up to 70%. These results question the validity of a number of fMRI studies and may have a large impact on the interpretation of weakly significant neuroimaging results."

The results showed that the packages correctly inferred statistical significance when considering independent voxels. However, when considering clusters of voxels as a single activation—as most activations are considered "clusters" or spatially correlated activations—the estimations of cluster sizes, smoothness, or the statistical threshold for what constitutes a "significantly activated" cluster were incorrect, leading to incorrectly large clusters.

The implication of their paper was that for the past fifteen years, these commonly used packages have overestimated activation extent. For well-designed studies with highly significant results, this would have virtually no effect on how results are interpreted. The same conclusions would likely have been made. Most studies' conclusions do not rely on absolute cluster size for their conclusions. Instead they draw conclusions based on the center of mass of activation, or whether a region was activated or not. Again, these studies would not be significantly affected.

Perhaps the greatest implications might be for pooled data in large data sets. If incorrectly large clusters are averaged across thousands of studies, then perhaps errors in interpretation of the extent of activation may crop up.

Unfortunately, the Eklund paper also went on to make a highly emotive statement: "These results question the validity of some 40,000 fMRI studies and may have a large impact on the interpretation of neuroimaging results." In a later correction, the authors removed the figure of 40,000 papers. Unfortunately, the media had picked up on "40,000" with the sensational statement that most fMRI studies are "wrong," which in itself is completely untrue by most definitions of the word "wrong." The studies may have simply slightly overestimated the size of activation. If a paper relied on this error to reach a conclusion, it was naturally treading on thin statistical ice in the first place and would likely be in question anyway. Most activation (and most of the voxels in the clusters found) are highly significant.

After many popular articles raised concern in the public, the issue eventually died down. Most of the scientists in the field who understood the issue were completely unfazed because the slightly larger clusters did not have any influence on the conclusions drawn by their papers or papers that they relied on for guiding their work.

This brings up an issue related to statistical tests. All statistical tests involve estimates on the nature of the signal and noise, so by definition are always somewhat "wrong." They are close enough, however. So much more goes into the proper interpretation of results. Most seasoned fMRI practitioners have gained experience in not over-interpreting aspects of the results that are not extremely robust.

While the Eklund paper has performed a service to the field by alerting it to a highly propagated error, it also raised false concerns about the high level of robustness and reproducibility of fMRI in drawing inferences about brain activation. Yes, the extent of brain activation has been slightly overestimated, but no, most of the papers produced using these analyses need to change in any manner, their conclusions.

#25: The Issue of Reproducibility

While fMRI is a robust and reproducible method, there is room for improvement. In science in general, it's been suggested that up to 70% of published results have failed to be reproduced. The reasons for this high fraction may be in part related to what we mean by "successfully reproduced" as well as pressure to publish findings that push the limits of what may be a reasonable level of interpretation. Some might argue that this number is an indication of the health of the scientific process by which a study's result either lives or dies based on whether it is replicated. If a study is not able to be replicated, the conclusions generally fade away from acceptance.

One researcher, Russ Poldrack of Stanford University, is spearheading an effort to increase transparency and reproducibility of fMRI data as he heads the Stanford Center for Reproducible Science. The goal is to increase further the reproducibility of fMRI and therefore move the field toward a fuller realization of its potential and less wasted work in the long run. He specifically is encouraging more replication papers, more shared data and code, more "data papers" contributing a valuable data set that may be reanalyzed, and an increased number of "registered studies" where the hypotheses and methods are stated up front before any data are collected. All of these approaches will make the entire process of doing science with fMRI much more transparent and able to be better leveraged by a growing number of researchers as they develop better processing methods or generate more penetrating hypotheses.

#26: Dynamic Connectivity Changes

The most recent "controversy" in the field of fMRI revolves around the interpretation of resting state scans. Typically, in the past, maps of resting state connectivity were created from entire time series lasting up to thirty minutes. The assumption was that brain connectivity remained somewhat constant over these long periods. There is evidence that this assumption is not correct. In fact, it's been shown that the connectivity profile of the brain changes fluidly over time. Today, fMRI practitioners now regard individual fMRI scans as rich 4D data sets with meaningful functional connectivity dynamics (dFC) requiring updated models able to accommodate this additional time-varying dimension. For example, individual scans today are often described in terms of a limited set of recurring, short-duration (tens of seconds), whole-brain functional connectivity (FC) configurations named "FC states." Metrics describing their dwell times, ordering, and frequency of transitions can then be used to quantify different aspects of empirically observed dFC. Many questions remain about the etiology of empirically observed FC dynamics as well as the ability of models, such as FC states, to accurately capture behavioral, cognitive, and clinically relevant dynamic phenomena.

Despite reports of dFC in resting humans, macaques, and rodents, a consensus does not exist regarding the underlying meaning or significance of dFC while at rest. Those who hypothesize it to be neuronally relevant have explored resting dFC in the context of consciousness, development, and clinical disorders. These studies have shown how the complexity of dFC decreases as consciousness levels decline; how dynamic interregional interactions can be used to predict brain maturity; and how dFC derivatives (e.g., dwell times) can be diagnostically informative for conditions such as schizophrenia, mild cognitive impairment, and autism.

Yet, many fMRI practitioners and researchers have raised concerns regarding the ability of current dFC estimation methods to capture neuronally relevant dFC at rest. These concerns include lack of appropriate null models to discern real dynamics from sampling variability, improper pre-processing leading to spurious dynamics, and excessive temporal smoothing (a real concern for sliding window techniques used to estimate FC states) that hinder our ability to capture sharp and rapid transitions of interest.[54] Finally, some have even stated that resting dFC is primarily a manifestation of sampling variability, residual head-motion artifacts, and fluctuations in sleep state—and as such, mostly irrelevant.

One reason for such discrepant views is that it is challenging to demonstrate the potential cognitive correlates of resting dFC, given the unconstrained cognitive nature of rest and scarcity of methods to infer the cognitive correlates of whole-brain FC patterns. When subjects are instructed to quietly rest, retrospective reports demonstrate that subjects often engage in a succession of self-paced cognitive processes including inner speech, musical experience, visual imagery, episodic memory recall, future planning, mental manipulation of numbers, and periods of heightened somatosensory sensations. Reconfigurations of FC patterns during rest could, to some extent, be a manifestation of this flow of covert self-paced cognition, even if other factors also contribute, such as random exploration of cognitive architectures and fluctuations in autonomic system activity and arousal levels.

Research is ongoing in fMRI to determine the neural correlates of dynamic connectivity changes, to determine the best methods for extracting this rich data, and to determine how this information may be used to better understand ongoing cognition in healthy and clinical populations and individuals.

In addition to those addressed in this chapter, other interesting controversies also come to mind—such as BOLD activation in white matter; BOLD tensor mapping;

the problem of reverse inference from fMRI data; the difference between mapping connectivity (ahem . . . correlation) and mapping fMRI magnitude; inferring brain region causation from different brain regions during activation using fMRI; and other non-BOLD contrast mechanisms such as temperature or elasticity. Challenges also come to mind—including the robust extraction of individual similarities and differences from fMRI data; the problem of how best to parcellate brains; the creation of fMRI-based biomarkers and potential utility of gradient coils.

The field of fMRI has advanced and further defined itself through many of these controversies. In my opinion, these indicate that the field is still growing and is quite robust and dynamic. It's important to keep in mind that in the end, consensus generally is reached—thus pushing our understanding forward. Controversies are healthy and welcomed. Let's keep them coming!

CONCLUSION

Functional MRI is a powerful, noninvasive brain-imaging technology that sits squarely at the interface of domains such as physics, engineering, statistics, advanced signal processing, physiology, and neuroscience. It has the profound advantage of piggybacking on the thriving MRI scanner industry—allowing nearly every clinical MRI scanner to be used as a brain-function imaging device. Since its inception in 1991, fMRI has grown and spread explosively. Functional MRI now is performed by thousands of labs worldwide and has made a massive impact on the field of neuroscience.

Figure 17 illustrates what I like to call the four pillars of fMRI: technology, methodology, interpretation, and applications. These pillars are actively developing, interacting, and feeding off each other as the field of fMRI grows. Each pillar represents at least one field in itself and

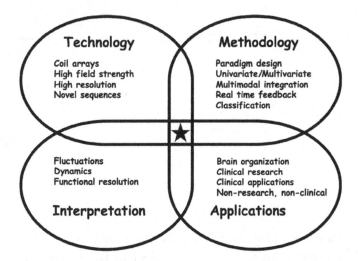

Figure 17 Each of the four "pillars" of functional MRI represents a key area where fMRI has grown. The four interact and evolve as fMRI evolves. The best science, depicted as the star in the middle, usually occurs in work that utilizes the best aspects of each pillar.

each depends on, pushes, and adapts to the other fields as they advance. The best fMRI work makes use of the latest advances in each of the four areas, engaging the best methods, technology, and understanding of the signal toward the applications that lend themselves most to what can be done.

This book has covered the basics and some of the advanced aspects of fMRI methodology. My goal was to provide a solid introduction, some advanced details and

By all measures, fMRI is a thriving, mature technology that continues to advance at an accelerating rate. It has considerable untapped potential to make further inroads, especially in clinical settings that will help us to more deeply understand the brain. The future is bright.

elements, and an overall useful perspective on all aspects of fMRI for a range of readers from the curious layperson to the sophisticated developer and user. I've attempted to define all essential technical terms and to spell out the basic concepts while simultaneously providing insight into some of the more salient and popular aspects of fMRI.

By all measures, fMRI is a thriving, mature technology that continues to advance at an accelerating rate. It has considerable untapped potential to make further inroads, especially in clinical settings that will help us to more deeply understand the brain. The future is bright.

GLOSSARY

Arterial spin labeling (ASL)
This is a pulse sequence used to map brain perfusion. It involves an initial inversion pulse—typically outside the imaging plane of interest—to "label" or "tag" inflowing blood. After a waiting period to allow blood to flow into the imaging plane and influence the magnetization, images are collected. Image collection alternates between applied label and no labeling applied. The pairs of images are averaged and subtracted to reveal perfusion.

B_0
The primary magnetic field in magnetic resonance imaging. Its strength typically is measured in Tesla (T). Typical field strengths are 1.5T to 3T, although scanners at 7T are becoming more popular. The highest field strength for human imaging is 11.7T at the National Institutes of Health. The MRI signal is directly proportional to the primary magnetic field.

Blocked design
A type of brain activation paradigm where long periods (>10 sec) of rest are interspersed with long periods of activation. These "blocks" typically repeat multiple times over a time series.

BOLD contrast
Blood oxygen level–dependent contrast. It was first coined by Seiji Ogawa and is the basis of most fMRI contrast. It is based on the fact that oxygenated blood has the same susceptibility as surrounding plasma and tissue, but deoxygenated blood has a lower susceptibility than surrounding plasma and tissue, causing magnetic field distortions and spin dephasing and thus a reduced signal. During activation, blood oxygenation locally increases, causing less spin dephasing and an increased signal.

Brain activation paradigm
A sequence of stimuli or tasks that the subject receives or performs in the scanner while undergoing fMRI. These sequences are played out during the collection of a time series of images.

Capillary

The smallest vessel in the vascular tree. Capillaries are typically two to ten microns in diameter and constitute about 2%–5% of the brain's total volume.

Cerebrospinal fluid (CSF)

The fluid that exists in the brain ventricles and sulci as well as within the spinal column.

Clustering

A processing approach by which the statistical threshold on individual voxels is lowered depending on the spatial proximity to other significant or nearly significant voxels.

$CMRO_2$

The cerebral metabolic rate of oxygen is the rate at which the brain consumes oxygen.

CO_2

Carbon dioxide: a common gas that, in the context of fMRI, is inhaled to cause vasodilation without causing brain activation. Vasodilation in turn causes an increase in blood flow and blood volume. It has been used to measure the potential for BOLD changes and to calibrate BOLD toward better localization and quantitative measures of activation-induced changes in $CMRO_2$.

Connectivity

In the context of fMRI, "connectivity" is typically defined as the degree to which the fMRI time series signal shows a temporal correlation. It is a problematic term since many factors may influence temporal correlation other than the degree of connectivity. In addition, factors may influence connectivity without influencing temporal correlation of the signal. Nevertheless, "connectivity" is now used interchangeably with fMRI time series correlation.

Correlation

The degree of temporal similarity between time series signals or the degree of spatial similarity between spatial patterns of activation.

Cortex

The layer of gray matter on the outermost surface of the brain. It is along this sheet that most cognitive and sensory-related neuronal activity takes place.

Cortical layer

A typical cortical ribbon contains several layers from the surface to the interface with white matter. Each layer consists of specific types and densities of neurons. It is known that some layers receive input and other layers send output to other regions.

dB/dT

This is the rate of magnetic field change over time. It commonly is brought up as a major limit to how quickly gradients can switch, because above a certain threshold the subject experiences muscle twitching in the nose, shoulders, or back.

Decoding

In the context of brain imaging, decoding is the operation by which an algorithm is able to determine a highly specific brain activity, with an accuracy greater than chance. Typically, a large number of training sets are analyzed by the algorithm until it learns to recognize the unique temporal and/or spatial pattern of activity that is associated with a specific stimulus or task.

Default mode network (DMN)

One of the most readily found networks in resting state fMRI, DMN is known to be more active during resting state and deactivated during externally driven attention or during cognitive tasks. The specific role of this network remains incompletely understood. The regions of the default mode network includes the posterior cingulate cortex, precuneus, medial prefrontal cortex, and angular gyrus.

Deoxyhemoglobin

The form of hemoglobin without oxygen. Hemoglobin has four heme units that bind to oxygen. If they are unbound, the molecule is paramagnetic and called deoxyhemoglobin. Deoxyhemoglobin distorts magnetic fields and causes signal attenuation or NMR phase shifts or both.

Dephasing

In the context of MRI, protons precess at specific frequencies. If protons (more generally called spins) are in a voxel and experience different magnetic fields, they precess at different frequencies and start to go out of phase—or "dephase." As the phase difference begins accumulating for more protons, they start to destructively add, and thus cancel their signals, causing a signal attenuation.

Diamagnetic
This property of materials causes them to repel applied magnetic fields. Most biological materials are diamagnetic.

Diffusion
The process by which particles such as water or gas molecules randomly move.

Diffusion tensor imaging (DTI)
An MRI-based technique in which the preferred directionality of diffusing spins is determined and typically mapped. This measurement makes use of directional diffusion gradients that encode diffusion as it projects on each orientation.

Dynamic connectivity
This refers to the change, over time, in connectivity or the temporal correlation.

Echo
An alignment of spin phase in the transverse plane that allows measurement of the signal. In a spin echo sequence, this occurs after a 180-degree refocusing pulse is applied. The time between the excitation pulse and the spin echo is exactly double the time between the excitation pulse and the application of the 180-degree refocusing or inversion pulse. In a gradient echo sequence, the echo is created during the free induction decay and is formed when the net gradient moment reaches zero. In other words, when the net area under the gradient waveforms adds to zero, the phase is at maximum and an echo forms.

Echo planar imaging (EPI)
An MRI pulse sequence in which an entire plane or slice of data is collected following a single RF excitation. It requires rapid gradient switching, but allows collection of an entire volume of MRI data in less than 2 sec. This is the most commonly used pulse sequence for fMRI, not only for its speed but also for the added temporal stability that comes with single-shot imaging.

Echo time (TE)

This is the time between the excitation pulse and the center of the readout gradient waveform during acquisition. This value determines the $T2^*$ and T2 contrast.

Electroencephalography (EEG)

This is a brain-mapping technique that measures voltage fluctuations on the surface of the scalp with electrodes and infers brain activity from these measurements. The voltage fluctuations are caused by the group summation of ionic currents within active neurons.

Encoding

In the context of brain imaging, encoding is the operation by which hypothesized temporal or spatial signatures of brain activation are used as regressors in order to determine the timing (if spatial signatures are used) or spatial location (if temporal signatures are used) of brain activity associated with a known stimulus or task.

Event-related fMRI

This is an activation paradigm that involves a very brief stimulus that elicits a brief hemodynamic response or an "impulse response." This may be the preferred activation paradigm in the field.

Excitation pulse

The radiofrequency pulse that moves the longitudinal magnetization into the transverse plane. Once it is in the transverse plane, the signal is detectable using receiver RF coils. Typically, an excitation pulse is a 90-degree pulse and is applied resonant frequency of the spins. It is typically accompanied by a "slice selective" gradient that causes only a plane to resonate at the RF excitation-pulse frequency.

Fludeoxyglucose (^{18}F-FDG)

A radioactive tracer used in positron emission tomography (PET) that is taken up by tissue in proportion to glucose metabolism.

fMRI adaptation

An activation paradigm that relies on habituation to identify voxels and regions selective to specific aspects of a stimulus. Identical or similar stimuli are presented sequentially with short intervals. Neurons that respond to the stimuli show habituation-related attenuation over time while those that do not are activated. This approach has been used to titrate selectivity to categories of stimuli.

Fourier transform
This mathematical operation decomposes a function of time or space into component frequencies. The inverse Fourier transform reverses this operation. Fast Fourier transform (FFT) algorithms work on discrete data sets that are powers of two. The fast Fourier transform, and inverse, is used in MRI for image reconstruction as it saves a considerable amount of time over a standard Fourier transform.

Free induction decay (FID)
The signal evolution, typically described by a single decaying exponential function, that occurs after the initial excitation RF pulse.

Frequency encoding
A generally unused and esoteric activation paradigm in which specific stimuli are temporally encoded by their presentation on-off frequency. Fourier analysis is performed on the time series to reveal each stimulus by its on-off frequency signature.

Functional connectivity MRI (fcMRI)
This is the name for the subfield of fMRI that characterizes the temporal correlation across voxels, regions, or parcels either during rest or during a task. Maps of connectivity may differ from maps of magnitude changes, as a connectivity modulation can occur without a magnitude change.

Functional magnetic resonance imaging (fMRI)
An MRI method by which a time series of blood flow, BOLD, or volume-sensitized images are collected during brain activation. A change in brain activity causes localized hemodynamic changes to induce small deviation of the signal that is detectible in the time series, allowing mapping of brain activation.

Gadolinium (Gd)
This is a chemical element that is used as a susceptibility or T1 contrast agent for mapping blood volume or vessel permeability with MRI. If injected as a bolus, it causes signal attenuation in proportion to blood volume. At a slower time constant, it changes T1 in relation to vascular permeability.

General linear model (GLM)
This is the framework by which multiple linear regression is carried out for fMRI analysis. The correlation of the data to multiple model functions is determined. Each model function is a hypothesized time course for a specific aspect of brain activity determined by the activation paradigm convolved with the hemodynamic response function.

Ghosting
In the MRI context, this is a common artifact in which some energy from the image is shifted a fraction (typically half) of an image.

Global signal regression
This is the removal of unwanted signal from the fMRI time series by regressing out the time series that is obtained by taking the average signal intensity across the entire brain at each time point—the global signal. This has been a controversial practice because while it does remove some spurious fluctuations, it may distort the final activation maps and in some cases change the sign of correlation values between regions. Use with caution.

Gradient coil
This is the piece of hardware consisting of wire windings that are configured to create—when current is passed through the windings—magnetic field gradients in typically three axes.

Gradient echo
A gradient echo occurs when, after an excitation pulse, the image-encoding gradient moment or the net area under the gradient strength and direction versus time plot is zero. The gradient echo is typically collected during the free induction decay.

Gray matter
The layer of brain that makes up the cortex and a substantial cross-section of the spinal cord. Relative to white matter, gray matter has more neurons and a greater variety of neurons that are packed closely together, as well as a much higher blood volume than white matter by up to a factor of 4.

Hemodynamic response function (HRF)
Neuronal activity causes a local hemodynamic response that is characterized by a function that shifts and spreads the neuronal activity event in time. A

brief neuronal activation will cause the hemodynamic response to start to increase after 2 sec, peak at about 5 sec, return to baseline by about 12 sec, and then exhibit a post-stimulus undershoot that can last between 10 and 40 sec. This function has been mathematically described as a Gamma Variate function or a difference between two exponentials. With all activation paradigms, this function is convolved with the known neuronal input timing. The assumption, valid for activation durations greater than 3 sec, is that the hemodynamic response is a linear function.

Hemoglobin
The molecule in red blood cells that carries and delivers oxygen to tissue.

Hz
Hertz: a measure of frequency. The units are cycles/sec.

Independent component analysis (ICA)
This is a commonly used post-processing approach in fMRI by which the signal is broken down into spatially independent components based on their time series patterns.

Inversion recovery
A pulse sequence that involves applying an inversion pulse (180-degree pulse) prior to the 90-degree excitation pulse. This allows for T1-weighted scans since the time for the flipped magnetization to recover is determined by T1. It can be used for perfusion imaging using ASL or for blood volume imaging using VASO.

K-space
The common term for the spatial frequency space where the raw MRI data is represented. Typically, k-space fills a grid in which the low spatial frequencies are in the center and high spatial frequencies are in the periphery. An inverse Fourier transform is performed on this data to convert it to an image—an operation typically known as image reconstruction.

Longitudinal relaxation
In MRI, longitudinal relaxation is the magnetization on the longitudinal axis. Once spins are excited by an RF pulse, the time to recover fully back to the longitudinal axis is determined by the tissue T1. Each tissue type has a different T1 value.

M₀

M_0

The net magnetization in the longitudinal plane. This is by definition 1 before excitation, and then grows back to 1 after excitation, as determined by T1. In the context of fMRI, it is the non-BOLD contribution to the signal change. BOLD changes T2 and T2* (transverse magnetization rates), but inflowing blood and possible proton density changes have an influence on the net magnetization independent of T2* or T2.

Magnetic resonance imaging (MRI)

This noninvasive imaging technique capitalizes on the observation that certain elements have precessing magnetization that is induced by a magnetic field. MRI measures magnetization of protons using excitation RF pulses applied at the resonance frequency at which they are precessing so that their magnetization is tipped in the transverse plane where it can be picked up by a receiver RF coil.

Magnetoencephalography (MEG)

This noninvasive technique uses multiple superconducting sensors to measure very small and rapid magnetic field fluctuations at the surface of the skull. Typically, a helmet of more than two hundred such sensors is used. From the magnetic field fluctuations, the sources of these fluctuations are inferred by mathematical modeling. However, source localization suffers from the inverse problem—which results from the fact that many possible source configurations can cause any given pattern of detected magnetic field on the skull.

Motion artifact

In the context of fMRI, these artifacts arise from head motion as well as chest wall motion. They typically manifest themselves near large spatial gradients signal intensity such as edges and signal-dropout regions. Chest wall motion causes field distortions that manifest themselves as changes in image ghosting or image warping. Many ways exist to correct motion artifacts but are far from perfect. Motion is still a problem in fMRI.

Multi-echo EPI

A technique by which multiple echoes are created and read out as imaging data for each RF excitation. Typically, in the context of fMRI, multi-echo EPI involves reading out two to five echoes during the FID following the RF excitation.

Multi-shot MRI
A common MRI approach in which the data for a single image or set of images is acquired using multiple RF excitations to fill in the raw data space or k-space. This method is used for most structural images in MRI. However, it is not used in echo planar imaging as it is less stable and considerably slower. The advantage is that the possible image resolution achievable is much higher than with single-shot approaches like EPI.

Multivariate analysis
In the context of fMRI processing it is the use of more than one measure to derive the statistical significance in drawing an inference about a task, a subject, or a population.

Myelin
The insulating sheath around nerve fibers that enables faster transmission of nerve impulses. Myelin tracts make up white matter and are found in the spinal cord. Myelin has also been found in primary sensory areas.

Naturalistic stimuli
Naturalistic stimuli draw on everyday continuous experiences. These may include movies, conversations, self-paced reading, listening to a story, performing a task such as driving, sorting objects, or playing a video game. Such stimuli have become more prominent in fMRI as post-processing methods for extracting and sorting salient neuronal information become more sophisticated.

Near-infrared spectroscopy (NIRS)
This is a method that relies on the observation that hemoglobin absorbs different wavelengths of light in the near-infrared range as a function of oxygenation. Localized changes in oxyhemoglobin, deoxyhemoglobin, and total hemoglobin can be measured over time using this method. Practically, the approach involves shining a light either through the skull or directly onto the brain and determining the oxygenation changes over time, based on the absorption spectrum changes.

Neuronal current imaging
A method that has yet to be convincingly demonstrated by which neuronal activation is directly visible to MRI. The electrical current that neurons generate produces very small and transient magnetic fields (on the order of 0.1nT).

These fields may cause spin-phase accumulation or spin dephasing, which in theory influence the MRI signal. So far no such transient changes have been demonstrated, likely because the effect is still too small by about an order of magnitude, given current time-series noise levels.

Neurovascular coupling
The spatial and temporal relationship between an increase in neuronal activity and subsequent blood volume, flow, and oxygenation changes.

Orthogonal
By definition, if two functions have a correlation of zero, they are orthogonal. In the context of fMRI, it is useful to design activation paradigms such that specific tasks eliciting activation in different brain regions are orthogonal to each other to ensure minimal mixing of task effects in post-processing. If two functions are not orthogonal, they can be mathematically "orthogonalized" at a cost of statistical power.

Oxyhemoglobin
The form of hemoglobin that is bound to oxygen. Hemoglobin has four heme units that bind to oxygen. If they are bound, the molecule is diamagnetic, like the rest of biologic tissue, and does not cause any phase shifts or attenuate the MRI signal.

Paramagnetic
The property of materials that causes them, in the presence of a magnetic field, to become magnetized in such a manner that attracts the fields. Deoxyhemoglobin and gadolinium are paramagnetic.

Phase
In the context of MRI, phase is the position in time of a signal that is oscillating. Spins in MRI have a specific frequency and phase. If they all experience the same frequency and have been synchronized by an RF pulse, they are in phase. If they begin to experience different magnetic fields, they start to precess at different frequencies and fall out of phase.

Phase encoding
In the context of MRI, spins are spatially encoded using gradients that alter their frequency at specific times, and therefore alter their phase as well. In k-space, spatial encoding based on the relative phase of spins is conventionally

depicted on the y-axis while spatial encoding based on the frequency differential is depicted on the x-axis.

In the context of fMRI paradigm design, phase encoding is a popular and powerful method for performing retinotopy or tonotopy in which a stimulus property is varied continuously and in a cyclic manner. This continuous change causes a corresponding continuous change in the spatial location of activation. Activation maps can then be created that show the activation corresponding to the phase of the stimulus modulation. This has been used in mapping regions that have a continuous spatial layout such as the visual, auditory, and somatosensory cortex.

Physiologic noise
Also known as physiologic fluctuations, this noise currently sets the upper limit on the temporal signal to noise ratio in fMRI time series at about 120 to 1. Components of physiologic noise are respiration, cardiac pulsation, vasomotion not related to neuronal activity, and movement. After a quarter century of trying, the field is still unable to eliminate or even substantially reduce physiologic noise.

Pial veins
The large veins found at the surface of the cortex and spinal cord that receive blood from capillaries. These are a major confound in attempting to resolve functional columns and layers as they contribute significantly to BOLD contrast and can overwhelm the subtler spatially specific changes in capillary and small-vein oxygenation that occur locally to specific layers and columns.

Positron emission tomography (PET)
PET is a method to image brain metabolism, neurotransmitter uptake, and blood flow by which compounds are labeled with positron-emitting radionuclides and injected into the body, acting as tracers. These are deposited in various amounts in the body. Opposing radiation coincidence detectors are able to localize the deposited isotopes. Typically rings of detectors in a cylinder configuration are used.

Pre-processing
The processing step in fMRI that typically involves motion correction, image registration, and slice-timing correction. Other steps may involve temporal and spatial smoothing as well as nuisance regression.

Radiofrequency (RF) coil

This is the coil that provides a resonant radiofrequency pulse to excite spins. Often the same coil is used to detect the projection of the spin magnetization on the transverse plane. Most commonly, a large head coil is used to provide a uniform excitation and an array of up to sixty-four receiver coils detect the signal as the sensitivity is increased with decreasing size.

Real time fMRI

This approach to fMRI involves displaying updated functional information (maps and/or time courses) in real time as the scan is being performed. A growing application is for neurofeedback in which functional activation information is fed back to the subject as they attempt to modify their brain activation.

Reconstruction

The performance of an inverse Fourier transform on the raw data from the scanner to create an MR image. Typically, this is performed automatically at the scanner by vendor-proprietary code.

Repetition time (TR)

In the context of fMRI using single-shot EPI, this is the time between each volumetric collection of data. In multi-shot pulse sequences, it is the time between each excitation pulse.

Resting state fMRI (rsfMRI)

This is a major area of fMRI where time series are collected without having the subject perform any task other than perhaps staring at a fixation point. From the time series data, the temporal correlations between voxels, multivoxel parcels, or regions are calculated. The source of this signal is generally understood to be spontaneous neuronal firing that is synchronized between functionally connected regions (i.e., left and right motor cortex).

SENSitivity Encoding (SENSE)

This is an approach to MR imaging that uses the sensitivity profile of multiple receiver RF coils to assist in spatial encoding. This approach allows either significantly shorter readout windows or higher in-plane resolutions for a given readout window duration. The advantages of a shorter readout window in the context of fMRI include the ability to collect more echoes during the free induction decay, and less distortion in the images collected with a shorter readout window. This technique also enables significantly higher in-plane resolution than is possible with a standard single-shot approach due to constraints of T2* decay and the biologic limits of gradient switching.

Simultaneous multi-slice (SMS) imaging
Also known as multi-band imaging, SMS is a recent advancement in MRI in which RF pulses of different frequencies are multiplexed in one pulse to excite multiple parallel planes at the same time. This speeds up the rate by up to a factor of 8 at which an imaging volume can be obtained.

Single-shot MRI
An MR imaging approach in which the raw data necessary to create a single plane, or in extreme cases, volume of data is collected with a single RF excitation pulse or set of RF pulses in the case of single-shot spin echo imaging. EPI is the most common type of single-shot approach.

Slice-timing correction
Because it can take up to 4 sec to collect an entire volume of data, a common pre-processing approach is to temporally shift time courses from slices collected at later times by the timing offset of the slice acquisition and then interpolating the signal to the time of the first slice collected in the volume. This helps when analyzing transient signals across slices collected at different times.

Smoothing
This is the act of averaging data across adjacent voxels or adjacent time points in pre-processing or sometimes in post-processing, in order to increase sensitivity. Often spatial smoothing is used prior to multi-brain normalization and averaging as the spatial transformation errors associated with template normalization are on a larger spatial scale than image resolution. Smoothing is generally beneficial as long as no interesting small spatial or temporal features are lost and averaged out in the process. It is generally safer to perform temporal smoothing since the HRF behaves as a low pass filter anyway; however, spatial smoothing is falling out of favor as extremely high-resolution maps and multivariate processing of voxel-wise patterns on single subjects are becoming more prevalent.

Spin

In the context of MRI, "spin" is a commonly used term for a proton precessing in a magnetic field.

Spin echo

A spin echo is formed by the combination of typically a 90-degree excitation pulse and then a 180-degree "refocusing pulse" that inverts the precession of spins and thus brings back into phase spins that have dephased following the 90-degree pulse. The moment in time that the spins are back in phase is called the spin echo. The time at which the spin echo occurs is double the time between the excitation pulse and the refocusing pulse.

Spiral imaging

In most echo planar imaging, the raw data in k-space is collected in a raster-like manner, starting at an edge and working back and forth until the entire image is filled in. With single-shot spiral imaging, the path through k-space is started at the center and then takes a spiral trajectory until the edge is reached. Generally, spiral k-space paths are more efficient, however off-resonance effects, rather than being manifest as a shift in location, are manifest as blurring. Pulsation artifacts, rather than being manifest as shifted lines that can be ignored, are manifest as ripples starting where the pulsation is occurring, interfering with the entire image. It is likely because of these drawbacks that spiral imaging has not caught on for fMRI. However, for multi-shot approaches to fMRI, it oversamples the center of k-space and therefore more effectively averages out shot-to shot-variability in the NMR phase, thus increasing stability.

Susceptibility

A property of all materials, susceptibility is the degree to which a material becomes magnetized in the presence of a magnetic field. Diamagnetic materials repel or diffuse applied magnetic fields and paramagnetic materials attract or concentrate applied magnetic fields.

T1

In MRI, T1 is the rate constant at which the magnetization exponentially returns to complete recovery of longitudinal magnetization. Different materials have different T1 values.

T2

In MRI, T2 is the rate constant by which the magnetization as measured using a spin echo sequence decays in the transverse plane. Different materials have different T2s, and the T2 of blood increases with oxygenation.

T2*

In MRI, T2* is the rate constant by which the magnetization as measured using a gradient echo sequence decays in the transverse plane. T2* is smaller than T2. Different materials have different T2* values, and T2* of blood increases with oxygenation.

Tesla (T)

A measurement of field strength, 1 Tesla = 10,000 Gauss. Typical MRI scanners are between 1.5 Tesla and 3 Tesla.

TI

This is the time between an inversion pulse and the excitation pulse of an inversion recovery pulse sequence. TI determines the degree of perfusion and/or longitudinal or T1 contrast.

Time series

In fMRI, this is the sequence of volumes that are collected over time as scanning is performed. Typically, a time series is between one minute and thirty minutes long, with an average time of about five minutes.

Transverse relaxation rate

This is the decay rate of signal in the transverse plane following an excitation pulse. It is characterized by T2 or T2*. It generally is much shorter than the longitudinal relaxation rate and is mostly due to spin-dephasing.

Vascular space occupancy (VASO)

In the context of fMRI, this is an approach that selectively nulls the blood signal based on the difference between blood T1 and gray matter T1. This allows imaging of blood volume change. With a blood volume increase the signal would decrease as more nulled blood fills each voxel. In this sequence, an inversion pulse is applied. Then, as the blood longitudinal relaxation is passing through the longitudinal magnetization null point, an excitation pulse is applied, thus only exciting nonblood spins. The contrast used by VASO has been shown to be more specific to small-vessel hemodynamic changes than BOLD contrast.

Vasodilation
This is the expansion of vascular sphincters allowing more arterial blood to pass through arteries and arterioles.

Voxel
The basic element in an image created using MRI or fMRI is the voxel. As a digital picture has pixels, a slab with a thickness contains voxels. A voxel is a 3D pixel.

White matter
The tissue in the brain and spinal cord that supports long-distance connections, white matter gets its pale appearance from the high levels of myelin enveloping nerve fiber tracts.

Xenon-CT
A method of computed tomography (CT) by which cerebral blood flow is assessed. The subject inhales Xenon gas, which travels into blood and tissue and is detected as a change in X-ray attenuation that is proportional to the Xenon concentration.

NOTES

Chapter 2

1. J. M. Harlow, "Passage of an Iron Rod through the Head. 1848," *Journal of Neuropsychiatry and Clinical Neuroscience* 11, no. 2 (1999): 281–283.

2. C. V. Rice, "Review of *Paul Lauterbur and the Invention of MRI*," *Journal of Chemical Education* 91, no. 5 (2014): 626–627.

3. D. Le Bihan, E. Breton, D. Lallemand, P. Grenier, E. Cabanis, and M. Laval-Jeantet, "MR Imaging of Intravoxel Incoherent Motions: Application to Diffusion and Perfusion in Neurologic Disorders," *Radiology* 161, no. 2 (1986): 401–407.

4. P. J. Basser, J. Mattiello, and D. LeBihan, "MR Diffusion Tensor Spectroscopy and Imaging," *Biophysical Journal* 66, no. 1 (1994): 259–267.

5. R. Xue, P. C. van Zijl, B. J. Crain, M. Solaiyappan, and S. Mori, "In Vivo Three-Dimensional Reconstruction of Rat Brain Axonal Projections by Diffusion Tensor Imaging," *Magnetic Resonance in Medicine* 42, no. 6 (1999): 1123–1127.

6. V. J. Wedeen, D. L. Rosene, R. Wang, G. Dai, F. Mortazavi, P. Hagmann, et al., "The Geometric Structure of the Brain Fiber Pathways," *Science* 335, no. 6076 (2012): 1628–1634.

7. D. K. Jones, ed., *Diffusion MRI* (New York: Oxford University Press, 2011).

8. M. E. Raichle and G. M. Shepherd, *Angelo Mosso's Circulation of Blood in the Human Brain* (Oxford: Oxford University Press, 2014).

9. S. S. Kety and C. F. Schmidt, "The Determination of Cerebral Blood Flow in Man by the Use of Nitrous Oxide in Low Concentrations," *American Journal of Physiology* 143 (1945): 53–66.

10. N. A. Lassen, D. H. Ingvar, and E. Skinhoj, "Brain Function and Blood Flow," *Scientific American* 239, no. 4 (1978): 62–71.

11. F. F. Jobsis, "Noninvasive, Infrared Monitoring of Cerebral and Myocardial Oxygen Sufficiency and Circulatory Parameters," *Science* 198, no. 4323 (1977): 1264–1267.

12. B. Chance, E. Anday, S. Nioka, S. Zhou, L. Hong, K. Worden, et al., "A Novel Method for Fast Imaging of Brain Function, Non-invasively, with Light," *Optics Express* 2, no. 10 (1998): 411–423.

13. E. Boto, A. Holmes, J. Leggett, G. Roberts, V. Shah, S. S. Meyer, et al., "Moving Magnetoencephalography towards Real-World Applications with a Wearable System," *Nature* 555 (2018): 657–661.

14. B. Chance, Y. Nakase, M. Bond, J. S. Leigh Jr., and G. McDonald, "Detection of 31P Nuclear Magnetic Resonance Signals in Brain by In Vivo and Freeze-Trapped Assays," *Proceedings of the National Academy of Sciences of the United States of America* 75, no. 10 (1978): 4925–4929.

15. J. J. Ackerman, T. H. Grove, G. G. Wong, D. G. Gadian, and G. K. Radda, "Mapping of Metabolites in Whole Animals by 31P NMR Using Surface Coils," *Nature* 283, no 5743 (1980): 167–170.

Chapter 3

1. R. A. Poldrack, T. O. Laumann, O. Koyej, B. Gregory, A. Hover, M. Y. Chen, et al., "Long-term Neural and Physiological Phenotyping of a Single Human," *Nature Communications* 6 (2015): 8885.

2. C. S. Roy and C. S. Sherrington, "On the Regulation of the Blood-supply of the Brain," *Journal of Physiology* 11 (1890): 85–108.

3. L. Pauling and C. D. Coryell, "The Magnetic Properties and Structure of Hemoglobin, Oxyhemoglobin and Carbonmonoxyhemoglobin," *Proceedings of the National Academy of Sciences of the United States of America* 22, no. 4 (1936): 210–216.

4. K. R. Thulborn, "My Starting Point: The Discovery of an NMR Method for Measuring Blood Oxygenation Using the Transverse Relaxation Time of Blood Water," *NeuroImage* 62, no. 2 (2012): 589–593.

5. P. T. Fox and M. E. Raichle, "Focal Physiological Uncoupling of Cerebral Blood Flow and Oxidative Metabolism during Somatosensory Stimulation in Human Subjects," *Proceedings of the National Academy of Sciences of the United States of America* 83 (1986): 1140–1144.

6. S. Ogawa, T. M. Lee, A. R. Kay, and D. W. Tank, "Brain Magnetic-Resonance-Imaging with Contrast Dependent on Blood Oxygenation," *Proceedings of the National Academy of Sciences of the United States of America* 87, no. 24 (1990): 9868–9872.

7. R. Turner, D. Lebihan, C. T. W. Moonen, D. Despres, and J. Frank, "Echo-Planar Time Course MRI of Cat Brain Oxygenation Changes," *Magnetic Resonance in Medicine* 22, no. 1 (1991): 159–166.

8. J. W. Belliveau, D. N. Kennedy, R. C. McKinstry, B. R. Buchbinder, R. M. Weisskoff, M. S. Cohen, et al., "Functional Mapping of the Human Visual-Cortex by Magnetic-Resonance-Imaging," *Science* 254, no. 5032 (1991): 716–719.

9. P. A. Bandettini, E. C. Wong, R. S. Hinks, R. S. Tikofsky, and J. S. Hyde, "Time Course EPI of Human Brain Function during Task Activation, *Magnetic Resonance in Medicine* 25, no. 2 (1992): 390–397; K. K. Kwong, J. W. Belliveau, D. A. Chesler, I. E. Goldberg, R. M. Weisskoff, B. P. Poncelet, et al., "Dynamic Magnetic-Resonance-Imaging of Human Brain Activity during Primary Sensory Stimulation, *Proceedings of the National Academy of Sciences of the United States of America* 89, no. 12 (1992): 5675–5679; S. Ogawa, D. W. Tank, R. Menon, J. M. Ellermann, S. G. Kim, H. Merkle, et al., "Intrinsic Signal Changes Accompanying Sensory Stimulation—Functional Brain Mapping with Magnetic-Resonance-Imaging," *Proceedings of the National Academy of Sciences of the United States of America* 89, no. 13 (1992): 5951–5955.

10. P. A. Bandettini, A. Jesmanowicz, E. C. Wong, and J. S. Hyde. "Processing Strategies for Time-Course Data Sets in Functional MRI of the Human Brain," *Magnetic Resonance in Medicine* 30, no. 2 (1993): 161–173.

11. B. Biswal, F. Z. Yetkin, V. M. Haughton, J. S. Hyde, "Functional Connectivity in the Motor Cortex of Resting Human Brain Using Echo-Planar MRI," *Magnetic Resonance in Medicine* 34, no. 4 (1995): 537–541.

12. E. S. Finn, X. Shen, D. Scheinost, M. D. Rosenberg, J. Huang, M. M. Chun, et al., "Functional Connectome Fingerprinting: Identifying Individuals Using Patterns of Brain Connectivity, *Nature Neuroscience* 18, no. 11 (2015): 1664–1671.

13. M. D. Greicius, B. Krasnow, A. L. Reiss, and V. Menon, "Functional Connectivity in the Resting Brain: A Network Analysis of the Default Mode Hypothesis," *Proceedings of the National Academy of Sciences of the United States of America* 100, no. 1 (2003): 253–258.

14. P. F. Liddle, "Is Disordered Cerebral Connectivity the Core Problem in Schizophrenia?," *NeuroScience News* 41, no 1 (2001): 62–73.

15. A. B. Waites, R. S. Briellmann, M. M. Saling, D. F. Abbott, and G. D. Jackson, "Functional Connectivity Networks Are Disrupted in Left Temporal Lobe Epilepsy," *Annals of Neurology* 59, no. 2 (2006): 335–343; S. J. Li, B. Biswal, Z. Li, R. Risinger, C. Rainey, J. K. Cho, et al., "Cocaine Administration Decreases Functional Connectivity in Human Primary Visual and Motor Cortex as Detected by Functional MRI," *Magnetic Resonance in Medicine* 43, no. 1 (2000): 45–51.

16. S. J. Li, B. Biswal, Z. Li, R. Risinger, C. Rainey, J. K. Cho, et al., "Cocaine Administration Decreases Functional Connectivity in Human Primary Visual and Motor Cortex as Detected by Functional MRI," *Magnetic Resonance in Medicine* 43, no. 1 (2000): 45–51.

17. A. Anand, Y. Li, Y. Wang, J. Wu, S. Gao, L. Bukhari, et al., "Activity and Connectivity of Brain Mood Regulating Circuit in Depression: A Functional Magnetic Resonance Study," *Biological Psychiatry* 57, no. 10 (2005): 1079–1088.

18. D. J. Watts and S. H. Strogatz, "Collective Dynamics of 'Small-World' Networks," *Nature* 393, no. 6684 (1998): 440–442.

19. D. Meunier, R. Lambiotte, and E. T. Bullmore, "Modular and Hierarchically Modular Organization of Brain Networks," *Frontiers in Neuroscience* 4 (2010): 200.

20. R. L. Buckner, J. Sepulcre, T. Talukdar, F. M. Krienen, H. Liu, T. Hedden, et al., "Cortical Hubs Revealed by Intrinsic Functional Connectivity: Mapping, Assessment of Stability, and Relation to Alzheimer's Disease," *Journal of Neuroscience* 29, no. 6 (2009): 1860–1873.

21. Y. Fan, F. Shi, J. K. Smith, W. Lin, J. H. Gilmore, and D. Shen, "Brain Anatomical Networks in Early Human Brain Development," *NeuroImage* 54, no. 3 (2011): 1862–1871.

22. K. Wu, Y. Taki, K. Sato, S. Kinomura, R. Goto, K. Okada, et al., "Age-Related Changes in Topological Organization of Structural Brain Networks in Healthy Individuals," *Human Brain Mapping* 33, no. 3 (2012): 552–568.

23. Z. Yao, Y. Zhang, L. Lin, Y. Zhou, C. Xu, and T. Jiang, "Abnormal Cortical Networks in Mild Cognitive Impairment and Alzheimer's Disease," *PLoS Computational Biology* 6, no. 11 (2010): e1001006.

24. R. M. Hutchison, T. Womelsdorf, E. A. Allen, P. A. Bandettini, V. D. Calhoun, M. Corbetta, et al., "Dynamic Functional Connectivity: Promise, Issues, and Interpretations," *NeuroImage* 80 (2013): 360–378.

25. M. S. Cetin, J. M. Houck, B. Rashid, O. Agacoglu, J. M. Stephen, J. Sui, et al., "Multimodal Classification of Schizophrenia Patients with MEG and fMRI Data Using Static and Dynamic Connectivity Measures," *Frontiers in Neuroscience* 10 (2016): 466.

26. C. Chang and G. H. Glover, "Time-Frequency Dynamics of Resting-State Brain Connectivity Measured with fMRI," *NeuroImage* 50, no. 1 (2010): 81–98.

Chapter 4

1. B. R. Rosen, J. W. Belliveau, J. M. Vevea, and T. J. Brady, "Perfusion Imaging with NMR Contrast Agents," *Magnetic Resonance in Medicine* 14, no. 2 (1990): 249–265.

2. J. W. Belliveau, D. N. Kennedy, R. C. McKinstry, B. R. Buchbinder, R. M. Weisskoff, M. S. Cohen, et al., "Functional Mapping of the Human Visual-Cortex by Magnetic-Resonance-Imaging," *Science* 254, no. 5032 (1991): 716–719.

3. K. R. Thulborn, J. C. Waterton, P. M. Matthews, and G. K. Radda, "Oxygenation Dependence of the Transverse Relaxation Time of Water Protons in Whole Blood at High Field," *Biochimica et Biophysica Acta* 714, no. 2 (1982): 265–270.

4. S. Ogawa, T. M. Lee, A. R. Kay, and D. W. Tank, "Brain Magnetic-Resonance-Imaging with Contrast Dependent on Blood Oxygenation," *Proceedings of the National Academy of Sciences of the United States of America* 87, 24 (1990): 9868–9872.

5. K. K. Kwong, J. W. Belliveau, D. A. Chesler, I. E. Goldberg, R. M. Weisskoff, B. P. Poncelet, et al., "Dynamic Magnetic-Resonance-Imaging of Human Brain Activity during Primary Sensory Stimulation," *Proceedings of the National Academy of Sciences of the United States of America* 89, no. 12 (1992): 5675–5679; S. Ogawa, D. W. Tank, R. Menon, J. M. Ellermann, S. G. Kim, H. Merkle, et al., "Intrinsic Signal Changes Accompanying Sensory Stimulation—Functional Brain Mapping with Magnetic-Resonance-Imaging," *Proceedings of the National Academy of Sciences of the United States of America* 89, no. 13 (1992): 5951–5955; P. A. Bandettini, E. C. Wong, R. S. Hinks, R. S. Tikofsky, and J. S. Hyde, "Time Course EPI of Human Brain Function during Task Activation," *Magnetic Resonance in Medicine* 25, no. 2 (1992): 390–397.

6. S. Ogawa, T. M. Lee, A. S. Nayak, and P. Glynn, "Oxygenation-Sensitive Contrast in Magnetic-Resonance Image of Rodent Brain at High Magnetic-Fields," *Magnetic Resonance in Medicine* 14, no. 1 (1990): 68–78.

7. R. S. Menon, S. Ogawa, J. P. Strupp, and K. Ugurbil. "Ocular Dominance in Human V1 Demonstrated by Functional Magnetic Resonance Imaging," *Journal of Neurophysiology* 77, no. 5 (1997): 2780–2787; K. Cheng, R. A. Waggoner, and K. Tanaka, "Human Ocular Dominance Columns as Revealed by High-Field Functional Magnetic Resonance Imaging," *Neuron* 32, no. 2 (2001): 359–374; E. Yacoub, N. Harel, and K. Uğurbil, "High-Field fMRI Unveils Orientation Columns in Humans," *Proceedings of the National Academy of Sciences of the United States of America* 105, no. 30 (2008): 10607–1061.

8. J. R. Polimeni, B. Fischl, D. N. Greve, and L. L. Wald, "Laminar Analysis of 7T BOLD Using an Imposed Spatial Activation Pattern in Human V1," *NeuroImage* 52, no. 4 (2010): 1334–1346.

9. R. S. Menon, D. C. Luknowsky, and J. S. Gati, "Mental Chronometry Using Latency-Resolved Functional MRI," *Proceedings of the National Academy of Sciences of the United States of America* 95, no. 18 (1998): 10902–10907.

10. N. K. Logothetis, J. Pauls, M. Augath, T. Trinath, and A. Oeltermann, "Neurophysiological Investigation of the Basis of the fMRI Signal," *Nature* 412, no. 6843 (2001): 150–157.

11. L. D. Lewis, K. Setsompop, B. R. Rosen, and J. R. Polimeni, "Fast fMRI Can Detect Oscillatory Neural Activity in Humans," *Proceedings of the National Academy of Sciences of the United States of America* 113, no. 43 (2016): E6679-E6685.

Chapter 5

1. M. NessAiver, "All You Really Need to Know about MRI Physics."

2. K. Setsompop, J. Cohen-Adad, B. A. Gagoski, T. Raij, A. Yendiki, B. Keil, et al., "Improving Diffusion MRI Using Simultaneous Multi-slice Echo Planar Imaging," *NeuroImage* 63, no. 1 (2012): 569–580.

3. D. A. Feinberg and K. Setsompop, "Ultra-fast MRI of the Human Brain with Simultaneous Multi-slice Imaging," *Journal of Magnetic Resonance* 229 (2013): 90–100; D. A. Feinberg and E. Yacoub, "The Rapid Development of High Speed, Resolution and Precision in fMRI," *NeuroImage* 62, no. 2 (2012): 720–725.

4. E. Yacoub, N. Harel, and K. Uğurbil, "High-Field fMRI Unveils Orientation Columns in Humans," *Proceedings of the National Academy of Sciences of the United States of America* 105, no. 30 (2008): 10607–10612.

5. F. H. Lin, K. W. Tsai, Y. H. Chu, T. Witzel, A. Nummenmaa, T. Raij, et al., "Ultrafast Inverse Imaging Techniques for fMRI," *NeuroImage* 62, no. 2 (2012): 699–705.

6. A. W. Song, E. C. Wong, and J. S. Hyde, "Echo-Volume Imaging," *Magnetic Resonance in Medicine* 32, no. 5 (1994): 668–671.

7. P. Kundu, N. D. Brenowitz, V. Voon, Y. Worbe, P. E. Vertes, S. J. Inati, et al., "Integrated Strategy for Improving Functional Connectivity Mapping Using Multiecho fMRI," *Proceedings of the National Academy of Sciences of the United States of America* 110, no. 40 (2013): 16187–16192; P. Kundu, S. J. Inati, J. W. Evans, W. M. Luh, and P. A. Bandettini, "Differentiating BOLD and Non-BOLD Signals in fMRI Time Series Using Multi-echo EPI," *NeuroImage* 60, no. 3 (2012):1759–1570.

8. J. Hennig, "Functional Spectroscopy to No-Gradient fMRI," NeuroImage 62, no. 2 (2012): 693–698.

9. D. K. Sodickson, M. A. Griswold, and P. M. Jakob, "SMASH Imaging," *Magnetic Resonance Imaging Clinics of North America* 7, no. 2 (1999): 237–254, vii–viii.

10. K. P. Pruessmann, M. Weiger, M. B. Scheidegger, and P. Boesiger, "SENSE: Sensitivity Encoding for Fast MRI," *Magnetic Resonance in Medicine* 42, no. 5 (1999): 952–962.

11. Feinberg and Setsompop, "Ultra-fast MRI of the Human Brain with Simultaneous Multi-slice Imaging."

12. M. F. Glasser, S. M. Smith, D. S. Marcus, J. L. Andersson, E. J. Auerbach, T. E. Behrens, et al., "The Human Connectome Project's Neuroimaging Approach," *Nature Neuroscience* 19, no. 9 (2016): 1175–1187.

Chapter 6

1. E. Amaro Jr. and G. J. Barker, "Study Design in fMRI: Basic Principles, *Brain and Cognition* 60, no. 3 (2006): 220–232.

2. S. M. Courtney, L. G. Ungerleider, K. Keil, and J. V. Haxby, "Object and Spatial Visual Working Memory Activate Separate Neural Systems in Human Cortex," *Cerebral Cortex* 6, no. 1 (1996): 39–49.

3. R. Birn, P. Bandettini, R. Cox, and R. Shaker, "Event-Related fMRI of Tasks Involving Brief Motion, *Human Brain Mapping* 7, no. 2 (1999): 106–114.

4. A. M. Blamire, S. Ogawa, K. Ugurbil, D. Rothman, G. McCarthy, J. M. Ellermann, et al., "Dynamic Mapping of the Human Visual-Cortex by High-Speed Magnetic-Resonance-Imaging," *Proceedings of the National Academy of Sciences of the United States of America* 89, no. 22 (1992): 11069–11073.

5. R. L. Buckner, P. A. Bandettini, K. M. O'Craven, R. L. Savoy, S. E. Petersen, M. E. Raichle, et al., "Detection of Cortical Activation during Averaged Single Trials of a Cognitive Task Using Functional Magnetic Resonance Imaging, *Proceedings of the National Academy of Sciences of the United States of America* 93, no. 25 (1996):14878–14883; G. McCarthy, M. Luby, J. Gore, and P. GoldmanRakic, "Infrequent Events Transiently Activate Human Prefrontal and Parietal Cortex as Measured by Functional MRI," *Journal of Neurophysiology* 77, no. 3 (1997): 1630–1634.

6. P. Bandettini and R. Cox, "Event-Related fMRI Contrast When Using Constant Interstimulus Interval: Theory and Experiment," *Magnetic Resonance in Medicine* 43, no. 4 (2000): 540–548.

7. R. M. Birn, R. W. Cox, and P. A. Bandettini, "Detection versus Estimation in Event-Related fMRI: Choosing the Optimal Stimulus Timing," *NeuroImage* 15, no. 1 (2002): 252–264.

8. S. A. Engel, D. E. Rumelhart, B. A. Wandell, A. T. Lee, G. H. Glover, E. J. Chichilnisky, et al., "fMRI of Human Visual-Cortex," *Nature* 369, no. 6481 (1994): 525.

9. K. Grill-Spector and R. Malach, "fMR-Adaptation: A Tool for Studying the Functional Properties of Human Cortical Neurons," *Acta Psychologica* 107, no. 1–3 (2001): 293–321.

10. A. G. Huth, T, Lee , S. Nishimoto, N. Y. Bilenko, A. T. Vu, and J. L. Gallant, "Decoding the Semantic Content of Natural Movies from Human Brain Activity," *Frontiers in Systems Neuroscience* 10 (2016): 81; A. G. Huth, W. A.

de Heer, T. L. Griffiths, F. E. Theunissen, and J. L. Gallant, "Natural Speech Reveals the Semantic Maps That Tile Human Cerebral Cortex," *Nature* 532, no. 7600, (2016): 453–458; A. G. Huth, S. Nishimoto, A. T. Vu, and J. L. Gallant, "A Continuous Semantic Space Describes the Representation of Thousands of Object and Action Categories across the Human Brain," *Neuron* 76, no. 6 (2012): 1210–1224; K. N. Kay, T. Naselaris, R. J. Prenger, and J. L. Gallant, "Identifying Natural Images from Human Brain Activity," *Nature* 452, no. 7185 (2008): 352–355.

11. C. Chu, Y. Ni, G. Tan, C. J. Saunders, and J. Ashburner, "Kernel Regression for fMRI Pattern Prediction," *NeuroImage* 56, no. 2 (2011): 662–673.

12. U. Hasson, Y. Nir, I. Levy, G. Fuhrmann, and R. Malach, "Intersubject Synchronization of Cortical Activity during Natural Vision," *Science* 303, no. 5664 (2004): 1634–1640; U. Hasson, O. Furman, D. Clark, Y. Dudai, and L. Davachi, "Enhanced Intersubject Correlations during Movie Viewing Correlate with Successful Episodic Encoding," *Neuron* 57, no. 3 (2008): 452–462.

13. B. Biswal, F. Z. Yetkin, V. M. Haughton, and J. S. Hyde, "Functional Connectivity in the Motor Cortex of Resting Human Brain Using Echo-Planar MRI," *Magnetic Resonance in Medicine* 34, no. 4 (1995): 537–541.

14. R. C. Craddock, G. A. James, P. E. Holtzheimer 3rd, X. P. Hu, and H. S. Mayberg, "A Whole Brain fMRI Atlas Generated via Spatially Constrained Spectral Clustering," *Human Brain Mapping* 33, no. 8 (2012): 1914–1928.

15. M. Bianciardi, M. Fukunaga, P. van Gelderen, S. G. Horovitz, J. A. de Zwart, K. Shmueli, et al., "Sources of Functional Magnetic Resonance Imaging Signal Fluctuations in the Human Brain at Rest: A 7 T Study," *Magnetic Resonance Imaging* (2009).

16. K. Murphy, R. M. Birn, D. A. Handwerker, T. B. Jones, and P. A. Bandettini, "The Impact of Global Signal Regression on Resting State Correlations: Are Anti-correlated Networks Introduced?," *NeuroImage* 44, no. 3 (2009): 893–905.

17. C. W. Wong, V. Olafsson, O. Tal, and T. T. Liu, "The Amplitude of the Resting-State fMRI Global Signal Is Related to EEG Vigilance Measures," *NeuroImage* 83 (2013): 983–990.

18. R. W. Cox, A. Jesmanowicz, and J. S. Hyde, "Real-Time Functional Magnetic-Resonance-Imaging," *Magnetic Resonance in Medicine* 33, no. 2 (1995): 230–236.

19. B. Sorger, B. Dahmen, J. Reithler, O. Gosseries, A. Maudoux, S. Laureys, et al., "Another Kind of 'BOLD Response': Answering Multiple-Choice Questions via Online Decoded Single-Trial Brain Signals," *Progress in Brain Research* 177 (2009): 275–292.

20. A. M. Owen, "Is Anybody in There?," *Scientific American* 310, no. 5 (2014): 52–57.

21. R. C. DeCharms, F. Maeda, G. H. Glover, D. Ludlow, J. M. Pauly, D. Soneji, et al., "Control over Brain Activation and Pain Learned by Using Real-Time Functional MRI," *Proceedings of the National Academy of Sciences of the United States of America* 102, no. 51 (2005): 18626–18631.

22. D. E. Linden, I. Habes, S. J. Johnston, S. Linden, R. Tatineni, L. Subramanian, et al., "Real-Time Self-Regulation of Emotion Networks in Patients with Depression," *PLOS ONE* 7, no. 6 (2012): e38115.

Chapter 7

1. J. V. Haxby, M. I. Gobbini, M. L. Furey, A. Ishai, J. L. Schouten, and P. Pietrini, "Distributed and Overlapping Representations of Faces and Objects in Ventral Temporal Cortex," *Science* 293, no. 5539 (2001): 2425–2430.

2. N. Kriegeskorte, R. Goebel, and P. Bandettini, "Information-Based Functional Brain Mapping," *Proceedings of the National Academy of Sciences of the United States of America* 103, no. 10 (2006): 3863–3868.

3. Y. Kamitani and F. Tong, "Decoding the Visual and Subjective Contents of the Human Brain," *Nature Neuroscience* 8, no. 5 (2005): 679–685.

4. N. Kriegeskorte, M. Mur, D. A. Ruff, R. Kiani, J. Bodurka, H. Esteky, et al., "Matching Categorical Object Representations in Inferior Temporal Cortex of Man and Monkey," *Neuron* 60, no. 6 (2008): 1126–1141.

5. A. G. Huth, S. Nishimoto, A. T. Vu, and J. L. Gallant, "A Continuous Semantic Space Describes the Representation of Thousands of Object and Action Categories across the Human Brain," *Neuron* 76, no. 6 (2012): 1210–1224.

6. A. Shmuel and D. Leopold, "Neuronal Correlates of Spontaneous Fluctuations in fMRI Signals in Monkey Visual Cortex: Implications for Functional Connectivity at Rest," *Human Brain Mapping* 29 (2008): 751–761.

Chapter 8

1. C. S. Roy and C. S. Sherrington, "On the Regulation of the Blood-Supply of the Brain. *Journal of Physiology* 11 (1890): 85–108.

2. P. T. Fox and M. E. Raichle, "Focal Physiological Uncoupling of Cerebral Blood Flow and Oxidative Metabolism during Somatosensory Stimulation in Human Subjects," *Proceedings of the National Academy of Sciences of the United States of America* 83 (1986): 1140–1144.

3. P. T. Fox, "The Coupling Controversy," *NeuroImage* 62, no. 2 (2012): 594–601.

4. S. Ogawa, T. M. Lee, A. S. Nayak, and P. Glynn, "Oxygenation-Sensitive Contrast in Magnetic-Resonance Image of Rodent Brain at High Magnetic-Fields," *Magnetic Resonance in Medicine* 14, no. 1 (1990): 68–78.

5. R. S. Menon, "The Great Brain versus Vein Debate," *NeuroImage* 62, no. 2 (2012): 970–974.

6. S. Ogawa, D. W. Tank, R. Menon, J. M. Ellermann, S. G. Kim, H. Merkle, et al., "Intrinsic Signal Changes Accompanying Sensory Stimulation—Functional Brain Mapping with Magnetic-Resonance-Imaging," *Proceedings of the National Academy of Sciences of the United States of America* 89, no. 13 (1992): 5951–5955.

7. R. S. Menon, S. Ogawa, D. W. Tank, and K. Ugurbil, "Tesla Gradient Recalled Echo Characteristics of Photic Stimulation-Induced Signal Changes in the Human Primary Visual-Cortex," *Magnetic Resonance in Medicine* 30, no. 3 (1993): 380–386.

8. E. Yacoub, N. Harel, and K. Uğurbil, "High-Field fMRI Unveils Orientation Columns in Humans," *Proceedings of the National Academy of Sciences of the United States of America* 105, no. 30 (2008): 10607–10612.

9. L. Huber, D. A. Handwerker, D. C. Jangraw, G. Chen, A. Hall, C. Stuber, et al., "High-Resolution CBV-fMRI Allows Mapping of Laminar Activity and Connectivity of Cortical Input and Output in Human M1," *Neuron* 96, no. 6 (2017): 1253–1263 e7.

10. R. M. Birn and P. A. Bandettini, "The Effect of Stimulus Duty Cycle and 'Off' Duration on BOLD Response Linearity," *NeuroImage* 27, no. 1 (2005): 70–82.

11. N. K. Logothetis, J. Pauls, M. Augath, T. Trinath, and A. Oeltermann, "Neurophysiological Investigation of the Basis of the fMRI Signal," *Nature* 412, no. 6843 (2001): 150–157.

12. X. P. Hu, T. H. Le, and K. Ugurbil, "Evaluation of the Early Response in fMRI in Individual Subjects Using Short Stimulus Duration," *Magnetic Resonance in Medicine* 37, no. 6 (1997): 877–884; X. Hu and E. Yacoub, "The Story of the Initial Dip in fMRI," *NeuroImage* 62, no. 2 (2012): 1103–1108.

13. R. B. Buxton, "Dynamic Models of BOLD Contrast," *NeuroImage* 62, no. 2 (2012): 953–961.

14. P. C. van Zijl, J. Hua, and H. Lu, "The BOLD Post-Stimulus Undershoot, One of the Most Debated Issues in fMRI," *NeuroImage* 62, no. 2 (2012): 1092–1102.

15. A. Devor, P. Tian, N. Nishimura, I. C. Teng, E. M. Hillman, S. N. Narayanan, et al., "Suppressed Neuronal Activity and Concurrent Arteriolar

Vasoconstriction May Explain Negative Blood Oxygenation Level-Dependent Signal," *The Journal of Neuroscience* 27, no. 16 (2007): 4452–4459.

16. G. Krueger and C. Granziera, "The History and Role of Long Duration Stimulation in fMRI," *NeuroImage* 62, no. 2 (2012): 1051–1055.

17. J. Frahm, G. Kruger, K. D. Merboldt, and A. Kleinschmidt, "Dynamic Uncoupling and Recoupling of Perfusion and Oxidative Metabolism during Focal Brain Activation in Man, *Magnetic Resonance in Medicine* 35, no. 2 (1996): 143–148.

18. P. A. Bandettini, K. K. Kwong, T. L. Davis, R. B. H. Tootell, E. C. Wong, P. T. Fox, et al., "Characterization of Cerebral Blood Oxygenation and Flow Changes during Prolonged Brain Activation, *Human Brain Mapping* 5, no. 2 (1997): 93–109.

19. P. A. Bandettini, "The Temporal Resolution of Functional MRI," in *Functional MRI*, ed. C. Moonen and P. Bandettini, 205–220 (New York: Springer-Verlag, 1999); R. S. Menon, D. C. Luknowsky, and J. S. Gati, "Mental Chronometry Using Latency-Resolved Functional MRI," *Proceedings of the National Academy of Sciences of the United States of America* 95, no. 18 (1998):10902–10907; R. S. Menon, J. S. Gati, B. G. Goodyear, D. C. Luknowsky, and C. G. Thomas, "Spatial and Temporal Resolution of Functional Magnetic Resonance Imaging," *Biochemistry and Cell Biology* 76, no. 2–3 (1998): 560–571; P. S. F. Bellgowan, Z. S. Saad, and P. A. Bandettini, "Understanding Neural System Dynamics through Task Modulation and Measurement of Functional MRI Amplitude, Latency, and Width," *Proceedings of the National Academy of Sciences of the United States of America* 100, no. 3 (2003): 1415–1419.

20. M. Misaki, W. M. Luh, and P. A. Bandettini, "Accurate Decoding of Sub-TR Timing Differences in Stimulations of Sub-voxel Regions from Multi-voxel Response Patterns," *NeuroImage* 66 (2013): 623–633.

21. E. Formisano, D. E. J. Linden, F. Di Salle, L. Trojano, F. Esposito, A. T. Sack, et al., "Tracking the Mind's Image in the Brain I: Time-resolved fMRI during Visuospatila Mental Imagery," *Neuron* 35, no. 1 (2002): 185–194.

22. L. D. Lewis, K. Setsompop, B. R. Rosen, and J. R. Polimeni, "Fast fMRI Can Detect Oscillatory Neural Activity in Humans," *Proceedings of the National Academy of Sciences of the United States of America* 113, no. 43 (2016): E6679–E6685.

23. K. McKiernan, B. D'Angelo, J. K. Kucera-Thompson, J. Kaufman, and J. Binder, "Task-Induced Deactivation Correlates with Suspension of Task-Unrelated Thoughts: An fMRI Investigation," *Journal of Cognitive Neuroscience* (2002): 96.

24. R. L. Buckner, "The Serendipitous Discovery of the Brain's Default Network," *NeuroImage* 62, no. 2 (2012): 1137–1145.

25. A. Shmuel, E. Yacoub, J. Pfeuffer, P. F. Van de Moortele, G. Adriany, X. P. Hu, et al., "Sustained Negative BOLD, Blood Flow and Oxygen Consumption Response and Its Coupling to the Positive Response in the Human Brain," *Neuron* 36, 6 (2002): 1195–1210.

26. A. Shmuel and D. Leopold, "Neuronal Correlates of Spontaneous Fluctuations in fMRI Signals in Monkey Visual Cortex: Implications for Functional Connectivity at Rest," *Human Brain Mapping* 29 (2008): 751–761.

27. Y. Ma, M. A. Shaik, M. G. Kozberg, S. H. Kim, J. P. Portes, D. Timerman, et al., "Resting State Hemodynamics Are Spatiotemporally Coupled to Synchronized and Symmetric Neural Activity in Excitatory Neurons," *Proceedings of the National Academy of Sciences of the United States of America* 113, no. 52 (2016): E8463–E71.

28. S. M. Smith, T. E. Nichols, D. Vidaurre, A. M. Winkler, T. E. Behrens, M. F. Glasser, et al., "A Positive-Negative Mode of Population Covariation Links Brain Connectivity, Demographics and Behavior," *Nature Neuroscience* 18, no. 11 (2015): 1565–1567.

29. C. M. Bennett, A. A. Baird, M. B. Miller, and G. L. Wolford, "Neural Correlates of Interspecies Perspective Taking in the Post-mortem Atlantic Salmon: An Argument for Proper Multiple Comparisons Correction," presented at the 15th Annual Meeting of the Organization for Human Brain Mapping. San Francisco, CA, 2009.

30. J. Gonzalez-Castillo, Z. S. Saad, D. A. Handwerker, S. J. Inati, N. Brenowitz, and P. A. Bandettini, "Whole-Brain, Time-Locked Activation with Simple Tasks Revealed Using Massive Averaging and Model-Free Analysis," *Proceedings of the National Academy of Sciences of the United States of America* 109, no. 14 (2012): 5487–5492.

31. E. Vul, C. Harris, P. Winkielman, and H. Pashler, "Puzzlingly High Correlations in fMRI Studies of Emotion, Personality, and Social Cognition," *Perspectives on Psychological Science* 4, no. 3 (2009): 274–290.

32. M. D. Fox, D. Zhang, A. Z. Snyder, and M. E. Raichle, "The Global Signal and Observed Anticorrelated Resting State Brain Networks," *Journal of Neurophysiology* 101, no. 6 (2009): 3270–3283; M. D. Fox, A. Z. Snyder, J. L. Vincent, M. Corbetta, D. C. Van Essen, and M. E. Raichle, "The Human Brain Is Intrinsically Organized into Dynamic, Anticorrelated Functional Networks," *Proceedings of the National Academy of Sciences of the United States of America* 102, no. 27 (2005): 9673–9678.

33. K. Murphy, R. M. Birn, D. A. Handwerker, T. B. Jones, and P. A. Bandettini, "The Impact of Global Signal Regression on Resting State Correlations: Are Anti-correlated Networks Introduced?," *NeuroImage* 44, no. 3 (2009): 893–905.

34. Z. S. Saad, S. J. Gotts, K. Murphy, G. Chen, H. J. Jo, A. Martin, et al., "Trouble at Rest: How Correlation Patterns and Group Differences Become Distorted after Global Signal Regression," *Brain Connectivity* 2, no. 1 (2012): 25–32.

35. C. W. Wong, V. Olafsson, O. Tal, and T. T. Liu, "The Amplitude of the Resting-State fMRI Global Signal Is Related to EEG Vigilance Measures," *NeuroImage* 83 (2013): 983–990.

36. T. T. Liu, A. Nalci, and M. Falahpour, "The Global Signal in fMRI: Nuisance or Information?," *NeuroImage* 150 (2017): 213–229.

37. M. Chen, J. Han, X. Hu, X. Jiang, L. Guo, and T. Liu, "Survey of Encoding and Decoding of Visual Stimulus via FMRI: An Image Analysis Perspective," *Brain Imaging and Behavior* 8, no. 1 (2014): 7–23.

38. M. Misaki, W. M. Luh, and P. A. Bandettini, "The Effect of Spatial Smoothing on fMRI Decoding of Columnar-Level Organization with Linear Support Vector Machine," *Journal of Neuroscience Methods* 212, no. 2 (2013): 355–361.

39. Y. B. Sirotin and A. Das, "Anticipatory Haemodynamic Signals in Sensory Cortex Not Predicted by Local Neuronal Activity," *Nature* 457, no. 7228 (2009): 475–479.

40. S. D. Muthukumaraswamy and K. D. Singh, "Spatiotemporal Frequency Tuning of BOLD and Gamma Band MEG Responses Compared in Primary Visual Cortex," *NeuroImage* 40, no. 4 (2008): 1552–1560.

41. P. A. Bandettini, E. C. Wong, A. Jesmanowicz, R. S. Hinks, and J. S. Hyde, "Spin-Echo and Gradient-Echo EPI of Human Brain Activation Using BOLD Contrast: A Comparative Study at 1.5 T," *NMR in Biomedicine* 7, no. 1–2 (1994): 12–20.

42. E. Yacoub, A. Shmuel, J. Pfeuffer, P. F. Van De Moortele, G. Adriany, P. Andersen, et al., "Imaging Brain Function in Humans at 7 Tesla," *Magnetic Resonance in Medicine* 45, no. 4 (2001): 588–594; T. Q. Duong, E. Yacoub, G. Adriany, X. Hu, K. Ugurbil, J. T. Vaughan, et al., "High-Resolution, Spin-Echo BOLD, and CBF fMRI at 4 and 7 T." *Magnetic Resonance in Medicine* 48, no. 4 (2002): 589–593.

43. P. W. Stroman, V. Krause, K. L. Malisza, U. N. Frankenstein, and B. Tomanek, "Extravascular Proton-Density Changes as a Non-BOLD Component

of Contrast in fMRI of the Human Spinal Cord," *Magnetic Resonance in Medicine* 48, no. 1 (2002): 122–127.

44. P. Douek, R. Turner, J. Pekar, N. Patronas, and D. Le Bihan, "MR Color Mapping of Myelin Fiber Orientation," *Journal of Computer Assisted Tomography* 15, no. 6 (1991): 923–929.

45. D. Le Bihan, E. Breton, D. Lallemand, P. Grenier, E. Cabanis, and M. Laval-Jeantet, "MR Imaging of Intravoxel Incoherent Motions: Application to Diffusion and Perfusion in Neurologic Disorders," *Radiology* 161, no. 2 (1986): 401–407; D. Le Bihan, R. Turner, C. T. Moonen, and J. Pekar, "Imaging of Diffusion and Microcirculation with Gradient Sensitization: Design, Strategy, and Significance," *Journal of Magnetic Resonance Imaging* 1, no. 1 (1991): 7–28.

46. D. Le Bihan, S. I. Urayama, T. Aso, T. Hanakawa, and H. Fukuyama, "Direct and Fast Detection of Neuronal Activation in the Human Brain with Diffusion MRI," *Proceedings of the National Academy of Sciences of the United States of America* 103, no. 21 (2006): 8263–8268; S. Kohno, N. Sawamoto, S. I. Urayama, T. Aso, K. Aso, A. Seiyama, et al., "Water-Diffusion Slowdown in the Human Visual Cortex on Visual Stimulation Precedes Vascular Responses," *Journal of Cerebral Blood Flow and Metabolism* 29, no. 6 (2009): 1197–1207.

47. K. L. Miller, D. P. Bulte, H. Devlin, M. D. Robson, R. G. Wise, M. W. Woolrich, et al., "Evidence for a Vascular Contribution to Diffusion FMRI at High B Value," *Proceedings of the National Academy of Sciences of the United States of America* 104, no. 52 (2007): 20967–20972.

48. P. A. Bandettini, N. Petridou, and J. Bodurka, "Direct Detection of Neuronal Activity with MRI: Fantasy, Possibility, or Reality?," *Applied Magnetic Resonance* 29, no. 1 (2005): 65–88.

49. T. K. Truong, A. Avram, and A. W. Song, "Lorentz Effect Imaging of Ionic Currents in Solution," *Journal of Magnetic Resonance* 191, no. 1 (2008): 93–99.

50. G. T. Buracas, T. T. Liu, R. B. Buxton, L. R. Frank, and E. C. Wong, "Imaging Periodic Currents Using Alternating Balanced Steady-State Free Precession," *Magnetic Resonance in Medicine* 59, no. 1 (2008): 140–148; T. Witzel, F. H. Lin, B. R. Rosen, and L. L. Wald, "Stimulus-Induced Rotary Saturation (SIRS): A Potential Method for the Detection of Neuronal Currents with MRI," *NeuroImage* 42, no. 4 (2008): 1357–1365.

51. N. Ofen, S. Whitfield-Gabrieli, X. J. Chai, R. F. Schwarzlose, and J. D. Gabrieli, "Neural Correlates of Deception: Lying about Past Events and Personal Beliefs," *Social Cognitive and Affective Neuroscience* 12, no. 1 (2017): 116–127.

52. Z. Yang, Z. Huang, J. Gonzalez-Castillo, R. Dai, G. Northoff, and P. Bandettini, "Using fMRI to Decode True Thoughts Independent of Intention to Conceal," *NeuroImage* 99 (October 2014): 80–92.

53. A. Eklund, T. E. Nichols, and H. Knutsson, "Cluster Failure: Why fMRI Inferences for Spatial Extent Have Inflated False-Positive Rates," *Proceedings of the National Academy of Sciences of the United States of America* 113, no. 28 (2016): 7900–7905.

54. S. Shakil, C. H. Lee, and S. D. Keilholz, "Evaluation of Sliding Window Correlation Performance for Characterizing Dynamic Functional Connectivity and Brain States," *NeuroImage* 133 (2016): 111–128.

FURTHER READING

Bandettini, Peter, ed. "Twenty Years of Functional MRI: The Science and the Stories." *NeuroImage* 62 (2) (2012): 575–588.

Blijsterbosch, Janine, Stephen M. Smith, and Christian F. Beckmann. *An Introduction to Resting State fMRI Functional Connectivity*. Oxford: Oxford University Press, 2017.

Fornito, Alex, Andrew Zalesky, and Ed Bullmore. *Fundamentals of Brain Network Analysis*. Cambridge, MA: Academic Press, 2016.

Huettel, Scott A., Allen W. Song, and Gregory McCarthy. *Functional Magnetic Resonance Imaging*, 3rd ed. Sunderland, MA: Sinauer Associates, 2014.

Poldrack, Russell A. *The New Mind Readers: What Neuroimaging Can and Cannot Reveal about Our Thoughts*. Princeton: Princeton University Press, 2018.

Poldrack, Russell A., Jeannette A. Mumford, and Thomas E. Nichols. *Handbook of Functional MRI Data Analysis*. Cambridge: Cambridge University Press, 2011.

INDEX

PETER A. BANDETTINI is chief of the Section on Functional Imaging, director of the Functional MRI Facility, and director of the Center for Multimodality Neuroimaging at the NIH National Institute of Mental Health in Bethesda, Maryland. He conducted some of the first successful fMRI experiments as a PhD student and has devoted his career to the development of methods for functional MRI. He is the former editor-in-chief of the leading journal *Neuro-Image* and is the recipient of several awards including the Organization for Human Brain Mapping Young Investigator Award.